TRUST
The Winning Formula for Digital Leaders

A PRACTICAL GUIDE FOR COMPANIES ENGAGED IN DIGITAL TRANSFORMATION

TRUST
The Winning Formula
for Digital Leaders

A PRACTICAL GUIDE FOR COMPANIES
ENGAGED IN DIGITAL TRANSFORMATION

Paul Mugge

Haroon Abbu

Gerhard Gudergan

With a Foreword By

Gerald C. Kane

WWW.PATTERNSOFDIGITIZATION.COM

Book Cover Design by Nihal Abbu

Cover Image: Licensed from Shutterstock (2020)

Library of Congress Control Number: 2020925745

Paul Mugge, Haroon Abbu, Gerhard Gudergan

TRUST, The Winning Formula for Digital Leaders. A PRACTICAL GUIDE FOR COMPANIES ENGAGED IN DIGITAL TRANSFORMATION

ISBN: 978-1-7363784-1-0 (paperback)
ISBN: 978-1-7363784-2-7 (hardcover)
ISBN: 978-1-7363784-0-3 (ebook)

Dedicated to All Digital Leaders

Especially the following who shaped the contents of this book.

Chuck Sykes, CEO, Sykes Enterprises

Andera Gadeib, CEO, Dialego

Larry Blue, CEO, Bell & Howell

Robert Kallenberg, Director of Strategy and Organization, Porsche AG

Brandon Batten, Owner & Operator, Flying Farmer LLC

Marc Schlichtner, Principal Key Expert, Product, Portfolio & Innovation Management, Siemens Healthineers

Seth Kaufman, President & CEO, Moët Hennessy North America

Deborah Leff, former Global Leader and Industry CTO of Data Science and AI, IBM

Krishna Cheriath, VP, Head of Digital, Data and Analytics, Zoetis Inc.

Dominik Schlicht, CEO, Talbot New Energy AG

Craig Melrose, Executive Vice President, Digital Transformation Solutions, PTC

Dagmar Wirtz, CEO, 3WIN

Rahul C. Basole, Managing Director and Global Lead for Visual Data Science, Accenture AI

FOREWORD

By Gerald C. Kane, PhD

The book you hold in your hands, *TRUST: The Winning Formula for Digital Leaders*, brings home a key message. While technology is obviously a factor in digital transformation, perhaps a more important factor is something much more human—trust. The ability of some leaders to engender the trust of their employees allows them to accomplish a digital transformation much faster and at much less cost to the organization.

The book is about (and for) digital leaders, the people in charge of changing the course of their organizations. The authors have studied companies undergoing digital transformation and concluded that character and competency of their leaders differentiates Digitally Mature organizations from Digitally Developing organizations. They bring it all together with interview chapters from fifteen digital leaders on how they build trust.

The authors have the credibility to tackle this task. They have considerable industry experience and have spent years studying the perils, and triumphs, of companies engaged in digital transformation. They have seen these transformations first hand. More importantly, they have engaged directly with the digital leaders they have featured. They know what makes these leaders tick.

I agree with the authors, an organization's response to digital disruption should focus on people and processes—and not on technology. My own research shows that digital disruption is primarily about people and that effective digital transformation involves changes to organizational dynamics and how work gets done. A focus limited to selecting and implementing the right digital technologies is unlikely to lead to success. The best way to respond to digital disruption is by changing the company culture to be more agile, risk-tolerant, and experimental.

Digital disruption won't end anytime soon; the average worker will probably experience numerous waves of disruption during his or her career.

What, then, does digital transformation involve?

It Involves Leadership

Having effective leaders is among the most critical factors associated with effective digital transformation. While it is important that these leaders have strong digital literacy, being forward-looking and change-oriented are far more important characteristics. Interestingly, the majority of even the most Digitally Mature companies report needing more and better leaders to steer their organization into an uncertain digital future. The difference is that these advanced companies are far more likely to be taking active steps to develop those leaders than less advanced ones.

It Involves Talent Development

We have observed a similar trend with respect to digital talent. Almost all companies report needing more and better talent, but the Digitally Mature companies are actually doing something to develop that talent. My research shows that over 90% of respondents indicated that they need to update their skills at least yearly to keep up with digital trends, but fewer than half reported that their companies supported their efforts to keep up.

It Involves Culture

Digitally Mature companies are more likely to demonstrate a number of distinct cultural characteristics. They are more likely to be risk-tolerant, experimental, innovative, and collaborative. These companies more often report spending time, money, and energy on efforts to strengthen these aspects of their culture, suggesting a possibly widening gap between leaders and laggards.

It Involves Organization

Digitally Mature companies are more likely to organize around cross-functional teams, and these teams look different than those at less mature counterparts. They are more likely to have autonomy around how to meet business objectives and are more likely to be evaluated as a unit. These teams are often protected from the bureaucratic policies and procedures of legacy companies to allow them to be more agile.

It Involves Strategy

Perhaps the biggest factor associated with digital maturity is that the organization's leaders have a clear and coherent digital strategy, and they effectively communicate that strategy to employees. Again, this strategy may or may not involve implementing digital technologies at all but simply carving out a viable response to digital trends. For example, Best Buy has developed an effective strategy in the face of competition from Amazon by focusing on people, showrooms, and relationships—not technology.

This book should not be taken to mean that implementing new digital technologies and platforms are somehow not a vital part of effective digital transformation efforts; they almost certainly will be. The authors and I simply argue that an exclusive focus on the technological aspects of digital transformation overlooks one of the most important aspects—building trust across the organization, which produces the confidence to change their leadership, talent, culture, organization, and strategy—in one big, sometimes disruptive step!

Profile

Dr. Gerald C. (Jerry) Kane is a Professor of Information Systems and Faculty Director of the Edmund H. Shea, Jr. Center for Entrepreneurship at Boston College's Carroll School of Management. He is also a Senior Editor at MIS Quarterly. He teaches companies' responses to digital disruption to undergraduate, graduate, and executive education students worldwide.

His first book, published by MIT Press in April 2019, encapsulates five years of research into how companies are responding effectively to digital disruption and transformation. For this book, he surveyed over 20,000 executives worldwide and interviewed over 100 executives and thought leaders to determine the key factors associated with Digitally Mature companies. He concludes that an organization's response to digital disruption should focus on people and processes and not necessarily on technology. He speaks nationally and internationally to executive audiences on the characteristics of Digitally Mature companies, the steps necessary to become more mature, and how leaders need to adapt their own skills and careers to lead effectively in a digital world. His next book, *The Transformation Myth: Leading Your Organization Through an Uncertain World*, focuses on how companies adapt to acute disruptions like COVID-19.

ACKNOWLEDGEMENTS

This book is based on years of research conducted by two prestigious institutions at the center of industry/university collaboration: Center for Innovation Management Studies (CIMS) at North Carolina State University, USA, and FIR e.V. at RWTH Aachen University, Germany.

Center for Innovation Management Studies (CIMS) at NC State University, USA

The Center for Innovation Management Studies (CIMS) is a global virtual industry/university collaborative research center initiated and originally funded by the National Science Foundation in 1984. Located at NC State since 2000, CIMS is chartered as an official research center of the UNC system by the Board of Governors and is attached to the Poole College of Management at NC State University. CIMS serves as a leading authority on innovation management and innovation research, creating, synthesizing, and disseminating information on innovation management in and for modern businesses. CIMS is nationally recognized as the only center of its kind focused exclusively on how to manage innovation.

The CIMS network comprises industry members and other industry clients, academic researchers, and CIMS Industry and Academic Fellows. Over 35 years, more than 120 researchers have contributed their skills and expertise in support of CIMS-sponsored projects. In July of 2020, the Center's research focus on data analytics and digital leadership was merged with Poole College of Management's *Business Analytics Initiative*.

FIR e.V. at RWTH Aachen University, Germany

The Research Institute for Rationalization (FIR) e.V. was founded in 1953 at the RWTH Aachen University with the mission to establish and intensify research in the field of business organization in Germany. Today the institute helps companies; it also researches, qualifies, and teaches in the areas of production management, logistics, service management, information management, and business transformation. FIR has been awarded "Innovator of the Year" by the magazine *Brand Eins* in 2018 and 2019.

FIR's strategic focus, among other things, is digital transformation, including topics such as Industry 4.0 and Smart Services. The center has approximately 190 members from companies and institutions, as well as public and private citizens. The scientific work of FIR is enriched by a large portion of experimental research in its own innovation lab as well Demonstration Factory Aachen, a production facility operated under real-world conditions.

We want to give special thanks to the team at FIR e.V. for their support and contributions to the book—especially Ruben Conrad, Gerrit Hoeborn, and Alexander Kwiatkowski for contributing significantly, from the first draft to the final book you are reading now.

Ruben Conrad is head of the business transformation department at RWTH Aachen University. He joined the department after his studies in mechanical engineering at the RWTH Aachen University. His work experience spans three continents, Asia, North America, and Europe, where he worked with Bayer AG and the University of St. Gallen. He has worked on several research and industry projects with a focus on the design of new value creation models and management systems in various industries.

Gerrit Hoeborn holds degrees in business administration and mechanical engineering from RWTH Aachen University and Tsinghua University. His research focus is business transformation and digital business strategies. Gerrit is responsible for the ecosystem design section at FIR e.V. at RWTH Aachen University.

Alexander Kwiatkowski holds a degree in technical communication with a focus on mechanical engineering from RWTH Aachen University. His research focuses on topics such as digital leadership, culture, innovation, technologies, and business strategy.

TABLE OF CONTENTS

TABLE OF FIGURES

INTRODUCTION

"Trust is the highest form of human motivation. It brings out the very best in people." (1)

— Stephen M.R. Covey

What the Book Is About

This book is about, and for, digital leaders, i.e., the people in charge of changing the course of their organizations. It is our intention to understand what makes these people tick and by doing so offer helpful advice to those who may have been thrust into these positions. Digital leaders are not defined by their rank or position. While some digital leaders may be the CEOs of their firms, they may also be the heads of key business functions like operations and marketing, or perhaps they fill a key staff role, like the organization's Chief Data Officer. No matter their position, we feel these people are better defined by their ability to earn trust.

The book examines both the character and competency of proven digital leaders. While it is true that being proficient in the new technologies that drive digital transformation, e.g., artificial intelligence, Internet of Things, big data analytics, cloud computing,

> The enduring human traits of these leaders far outweigh their proficiency in the ever-evolving field of information technology.

1

and others, is definitely an advantage, it is our point of view that the enduring human traits of these leaders far outweigh their proficiency in the ever-evolving field of information technology. The ability of these leaders to engender the trust of their employees allows them to accomplish digital transformation much faster and at much less cost to the organization. The book is devoted to demonstrating this premise.

How the Book Is Organized

Although it may sound like one, *TRUST: The Winning Formula for Digital Leaders* is not some exposé on one of the most basic of human behaviors. Nor does it contain a lot of shortcuts on how—in a period of enormous economic and social change—digital leaders are expected to build and *keep* the trust of their employees. There is no quick fix for what we are talking about.

A Pandemic Hits

Here we should talk about the effect COVID-19 is having on these transformations—and their leaders. We began writing the book before this horrible disease started to spread. In fact, we were publishing articles about the impact of leadership on digital transformation in research journals almost coincident with this event. (2,3)

Then we had a decision to make… would we revise the book draft to focus on the pandemic, or just allow digital leaders to address its effects if they chose? We are certainly not epidemiologists, but we believe the pandemic will pass. And to strengthen our argument, we have over 60 years of research touting the benefits of trust in all types of relationships—personal, marital, organizational, and societal. (See Chapter 2, "Science of Trust.") We know from the literature that trust is enduring, and it pays big benefits.

> If people don't trust you—and you don't trust them—there is no reason for them to follow you anywhere.

The next big question COVID-19 raises is: Will the progress of digital leaders be thwarted or accelerated by the pandemic? This is a bigger issue that touches not only the leaders. As we will point out in Chapter 1, "Trust Is So Important," senior management of most organizations fear their organizations lack the skill—or even the will—to change themselves into digital businesses. Has this fear abated now that the whole company is perched on a burning platform? We doubt it.

During the pandemic, you've seen this whole scenario playing out in the stock market. The share price of digital companies—the ones that help people work, buy, educate, and play from home—have soared. Meanwhile, the price of other stocks (e.g., travel, leisure, restaurants, brick-and-mortar retail, etc.) have declined. Several of our digital leaders will address this issue. Maybe this book is more urgent for companies in these sectors now than it was pre-pandemic.

Anyway, while the pandemic is definitely a factor, we do think it will pass. So, let's return to what trust is and what it can buy you and your organization.

Back to the Book

Everybody knows that trust can take years to develop, seconds to destroy, and forever to repair. Establishing trust is just plain hard. In just about any given business publication, you will find a laundry list of skills, qualities, behaviors, and competencies that are attributed to successful leaders. Core competencies that bubble to the top are strategic thinking, effective communication, a desire to develop others, decision-making, creating a vision, ability to have tough conversations, and others. But here's the thing… if there's no trust, none of these other competencies even matter. If people don't trust you—and you don't demonstrate that you trust them—there is no reason for them to follow you anywhere.

> The interviews are the jewels of the book. They provide, better than we can, the first-hand experience of people who have walked in these moccasins.

This book is first and foremost a *handbook* for busy digital transformation leaders. It is meant to be a quick reference for why, for example, being open and honest with people is so important to your success, and ultimately theirs too. Consequently, we are offering the book in a number of formats—as a classic hardbound business book, in paperback, and as an e-book that you can read on Kindle or your smartphone. We designed the book to go with you wherever you go. You can take it to meetings; you can read it while waiting in airports or crammed into your airline seat; or you can delve into it late at night in the comfort of your home. Who said recasting an organization's business model was only a day job?

Within the framework of a conventional book, we include a set of in-depth interviews with thirteen proven digital leaders. The interviews are the jewels of the book. They provide, better than we can, the first-hand experience of people who have walked in these moccasins. We also have attached a number of findings from our rather extensive research into the topic of digital transformation. For those needing facts to back up our assertions, we offer these as well.

The first chapter, "Trust Is So Important," sets the stage for the purpose of the book. For the past several years, we have followed the research done by North Carolina State University, Poole College of Management's Enterprise Risk Management (ERM) Initiative with Protiviti, a public consulting firm. (4) Their findings, based on hundreds of corporate board officers, show an ever-increasing fear that their organizations are not equipped for digital transformation. Most recently, the results show that many of these organizations lack not only the skills but also the *will* to undertake changes of this size and scale. We think this situation is directly linked to the abilities (or lack of) of these organizations' digital leaders. In summary, this whole chapter builds the case that for organizations to have the will to transform themselves, they will need leaders who can rapidly build their trust.

Chapter 2, "Science of Trust," covers over 60 years of study by a series of prominent researchers in the field of trust. We show the considerable contributions of people like Roger Mayer, himself a professor at the Poole College of Management at NC State University. These researchers are, you might say, the pioneers of managerial trust. They built the foundation that other researchers now stand on. From these researchers, we see a definite pattern emerge. Trust is built on two foundations—*character* and *competence*. These two domains of trust work in tandem to produce the kind of digital leadership qualities we are seeking.

We introduce readers to Stephen M.R. Covey's powerful concept of the connections between trust and both the speed and cost of change: "When trust goes up, speed will increase, and costs will go down." (1) Could the lack of trust in digital leadership go so far as to impede digital transformation? We think it does. This is a powerful idea that we intend to explore through interviews with digital leaders.

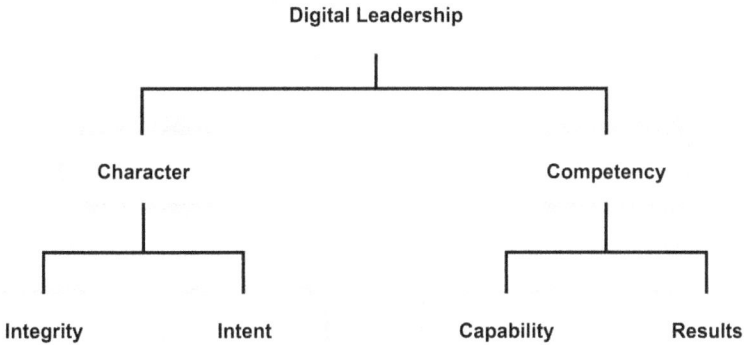

Figure 1. *Digital Leadership Model (Adapted from Covey's Trust Model)*

Covey's trust model takes these foundations a step further. In his popular book *The Speed of Trust,* Covey defines the Four Core Values of trust that underpin these two foundations:

1. *Integrity: Are you believable? (Character)*
2. *Intent: What is your agenda? (Character)*

3. *Capability: Are you relevant? (Competence)*
4. *Results: What is your track record? (Competence)*

We think you can see the important and self-effacing questions these core values raise. It is not easy to be a trustworthy digital leader.

In addition, Covey breaks these values down into a set of subfactors, e.g., *courage* and *humility* under the core value of *integrity*. That helps us examine a digital leader's behavior at a much more detailed—and more importantly—actionable level. Covey's "extended" model of trust is the basis of our interview guide.

Chapter 3, "Digital Maturity Demands Digital Leaders," explains how we used the Carnegie Mellon Maturity Index (CMMI) to measure the actions and behaviors of Digitally Developing companies compared to those who believe they are Digitally Mature. In an extensive global survey, called *Patterns of Digitization*, we investigated companies of all sizes and industries. (2) Again, the differences in leadership are quite pronounced. Digital leaders definitely "walk the talk." In our view, Digitally Mature leaders are more knowledgeable, exhibit entrepreneurial behaviors, foster timely and open dialogue, and through almost *relentless* communications promote digital transformation. We assert that the behaviors of these executives affect the implementation of digital transformation strategies as their actions directly impact employee performance.

> Digitally mature leaders are more knowledgeable, exhibit entrepreneurial behaviors, foster timely and open dialogue, and through almost relentless communications promote digital transformation.

The main point of this is that Digitally Mature organizations are managed by digital leaders. They are inseparable. Moreover, if you want to be a digital leader and understand what we have learned from them, please read the rest of the book.

Chapter 4, "Listening to Successful Leaders," sets up our interview process. In this chapter, we go into Covey's model in more depth and explain the real meaning behind the core values and their sub-values, like openness and honesty.

Most importantly, we share the 20-question interview guide we developed and explain the process we used for picking the leaders to be interviewed. Again, the interview is not a test; we know these people to be recognized digital leaders. We are trying to ascertain *how* they and their organizations have achieved their current position. Specifically, we want to know how their trust-inducing actions influenced the speed and cost of the organization's transformation.

Chapter 5, "The Pulse of Digital Leaders," captures the results of our interviews. We create a profile to introduce and position each interviewee, who was allowed to select the questions he or she is most passionate about. A few questions are common to all interviewees, like "What is your digital strategy and vision?" Each of the interview chapters is headlined by a key theme—in other words, the story behind the digital transformation taking place at these companies.

Importantly, through this process, we hope to gain the trust of the participants and emphasize the importance of speed for feeding this knowledge back to digital leaders. This is the reason for self-publishing this information and developing an e-book that readers can read on their mobile device!

Chapter 6, "The Winning Formula," is where it all comes together. Here we take a hard look at what these digital leaders have told us. What are the considerable issues facing the field of digital leadership? Before delving into the actions these leaders are taking, we thought it is important to describe the environment they work in. (In Chapter 6, see the section, "General Observations.") We then take a deep dive into the specific actions they are taking to set themselves and their organizations apart. We call this our "formula" for successful leaders to follow. Note the formula contains quite a list of actions. Also note, we call these actions, rather than beliefs or traits. Given the amount of almost frantic competition in every industry to transform their organizations into digital businesses, there is simply no time to just "grow into them." (In Chapter 6, see the section, "The Winning Formula: Fifteen Key Actions.")

We end the book with an interesting challenge: We invite you to test your own actions against the Winning Formula. Or your team can use the tool to test you! Sounds a bit risky doesn't it? But that's why we named the book *Trust: The Winning Formula for Digital Leaders*.

Opportunity or Threat?

George Westerman, principal research scientist with the MIT Sloan Initiative on the Digital Economy, asserts that Digitally Mature companies generate 9% higher revenue from their physical assets. (5) Andrew McAfee, co-director of the Initiative on the Digital Economy in the MIT Sloan School of Management, and his colleagues found that "data-driven companies are on average 5% more productive and 6% more profitable than other competitors in the market." (6) Further studies confirm that the so-called "digital masters" (the leaders we are writing about), are characterized by visionary management and digital capabilities and are 26% more profitable compared to their competitors. (6) We can all agree that digital transformation represents a significant change in the basic pattern of how organizations create value. Most importantly, we can see that properly equipped digital leaders do make a difference.

ERM and Protiviti researchers concluded that the question is not *if* digital is going to up-end their current business model, but rather *when*. Even when executives are aware of emerging technologies that have obviously disruptive potential, it is often difficult to have the vision or foresight to anticipate the nature and extent of change. Overcoming these hurdles represents the major challenges leadership teams confront when digitizing their businesses. However, there is no other choice, because the trend towards digital transformation will not weaken, but on the contrary, "The innovations and disruptions of the past ten years have been nothing short of astonishing... they're just the warm-up acts for what's to come." (1)

So why do organizations struggle so much with making this transition?

About Us

Our writing has been informed by our combined 75-plus years in industry.

This includes Paul Mugge's 35 years in product development, global business strategy, and business innovation services at IBM and 15 years of researching and teaching innovation as the Executive Director of the Center for Innovation Management at NC State University. It was through this experience that Mugge learned that more important than technology or product innovation is *business model* innovation. It, by far, is the hardest to copy—and the most important to establishing long-term growth. He captured these ideas—and the actual methods to perform business model innovation—in a book he co-authored with Dr. Stephen Markham, titled, *Traversing the Valley of Death: A practical guide for corporate innovation leaders*. Most importantly, Mugge spent his entire career – both at IBM and NC State University - listening and sharing what he has learned with scores of organizations that intend to separate themselves through innovation.

Haroon Abbu is Vice President of Digital, Data, and Analytics at Bell & Howell headquartered in Research Triangle Park, North Carolina. He is an accomplished leader with more than 20 years' industry experience in leading technology-enabled business transformation. Additionally, Haroon conducts research in the field of digital transformation in collaboration with the Center for Innovation Management Studies (CIMS) at NC State University and RWTH Aachen University, Germany. He presents regularly at leading industry conferences and has published his work in industry publications and academic journals such as the Journal of Business Research and Research-Technology Management.

Gerhard Gudergan is the deputy managing director at FIR Institute for Industrial Management at RWTH Aachen University in Germany. He is a renowned researcher with over 20 years of experience in the fields of service management, business transformation and digital leadership. Besides his research, Gudergan translates his entrepreneurial mindset into numerous spin-offs. Currently his focus is on the Metropolitan Cities Initiative, as the acting head of this latest project. The vision is to use urban innovation to transform the fifth largest metropolitan region in Europe, Rhine-Ruhr, into the most livable metropolitan area possible.

Chapter References

1. Covey, S.M.R. (2006). *The Speed of Trust: The one thing that changes everything.* Simon & Schuster.
2. Mugge, P., Abbu, H., Michaelis, T.L., Kwiatkowski, A., & Gudergan, G. (2020). Patterns of Digitization: A Practical Guide for Organizations Engaged in Digital Transformation. *Research - Technology Management*, 63(2), 27–35.
3. Abbu, H., Kwiatkowski, A., Mugge, P., Gudergan, G. (2020). *DIGITAL LEADERSHIP - Character and Competency Differentiates Digitally Mature Organizations.* IEEE International Conference on Engineering, Technology, and Innovation (ICE/ITMC).
4. NC State ERM Initiative and Protiviti. (2017). *Executive Perspectives on Top Risks for 2018: Key Issues Being Discussed in the Boardroom and C-Suite.*
5. Westerman, G., Bonnet, D., & McAfee, A. (2014). *Leading Digital: Turning technology into business transformation* (pp. 9–14). Harvard Business Review Press.
6. McAfee, A., Brynjolfsson, E., Davenport, T. H., Patil, D. J., & Barton, D. (2012). Big data: The management revolution. *Harvard Business Review, 90(10)*: 60–68.

CHAPTER 1

TRUST IS SO IMPORTANT

"Leaders establish trust with candor, transparency, and credit."

– Jack Welch

Before delving into the trust-building traits of digital leaders, let us first unpack what constitutes their chief task—leading digital transformation. We intend to break digital transformation down into several steps:

1. *Putting a premium on data,*
2. *Digitization of data,*
3. *Digitalizing business processes, and finally,*
4. *Digital transformation.*

Satish Nambisan, Professor of Technology Management at the Weatherhead School of Management, and his colleagues, in their *MIS Quarterly* article "Reinventing Innovation Management in a Digital World," characterize digital transformation as the creation of, and consequent change in, market offerings, business processes, or models that result from the use of digital technology. (1) In most instances, digital transformation represents a fundamental change in the organization's underlying mindset, systems, data, and tools needed to reposition its entire business design.

However, achieving a digital transformation can be quite a journey!

Unpacking Digital Transformation

We do not want to confuse you by introducing similar sounding terms like "digitization" and "digitalization," but this is not just a semantics game. Digital leaders need to know the difference. We will differentiate between the two. Moreover, we are going to demonstrate how many so-called digital leaders spend far too much time on digitization than they do on the digitalization of business processes, and ultimately digital transformation.

Digital transformation requires a much broader adoption of digital technology and cultural change than either digitization or digitalization. *Digital transformation is more about people than it is about digital technology.* It requires organizational changes that are customer-centric, backed by leadership, driven by radical challenges to corporate culture and the leveraging of technologies that empower and enable employees. Digital transformation requires trust.

Let us look into these steps.

Putting a Premium on Data

Ginni Rometty, former CEO of IBM, said data, particularly big data, "is like oil, and enterprises need help in extracting its value." Abhishek Mehta, founder of Tresata, expanded on Rometty's comment, "Just like oil was a natural resource powering the last industrial revolution, data is going to be the natural resource for this industrial revolution. Data is the core asset, and the core lubricant, for not just the entire economic models built around every single industry vertical but also the socioeconomic models."

We like Ms. Rometty's eloquent analogy to oil, but to have data be the core lubricant, you must still understand the steps of digitization and digitalization. These processes make data ubiquitous.

Digitization of Data

Wikipedia defines digitization as "the representation of an object, image, sound, document or signal (usually an analog signal) by generating a series of numbers that describe a discrete set of its points or samples. Simply speaking, digitizing simply means the conversion of analog source material into a numerical format." This is a powerful concept. Heck, it is the basis of the Internet of Things (IoT) technology—a mainstay of digital transformation and concepts like Industry 4.0 and Smart Manufacturing.

Digitization is of crucial importance to data processing, storage, and transmission, because it allows information of all kinds in all formats to be carried with the same efficiency and also intermingled. Though analog data is typically more stable, digital data can more easily be shared and accessed and can, in theory, be propagated indefinitely. This is why it is the favored way of preserving information for organizations worldwide.

Digitization largely refers to the *internal* optimization of processes (e.g., work automation, paper minimization) and results in cost reductions. Conversely, digitalization is a strategy or process that goes beyond the implementation of technology, implying a deeper, core change to the entire business model and the evolution of work.

Digitalization of Business Processes

Going back to the dictionary, *Gartner Glossary* does a good job of defining its meaning. It defines digitalization as "the use of digital technologies to change a business model and provide new revenue and value-producing opportunities; it is the process of *moving* to a digital business." Note the emphasis on "moving." Nevertheless, this is where real transformation starts. A digital business is the result of a multitude of digitalization processes and is an essential step *towards* digital transformation.

What does the digitalization of business processes accomplish?

Digitalization has several benefits. It immediately establishes the organization's digital presence. This is probably the most visible advantage. The presence on the Internet, through tools such as online stores, social networks, blogs, corporate pages, etc., multiplies the visibility of the company and sales channels. For some companies, this presence is the focus of their digital strategy, and they have even shifted their business from traditional forms to online commerce, with all that entails.

> Digitalization has several benefits. It immediately establishes the organization's digital presence. This is probably the most visible advantage.

It also enables even better decision-making. The digitalization of business makes it possible to have continuous contact with the customer, and this allows us to get to know him or her better. Some companies go further and apply big data—data on steroids, if you will—when making all kinds of decisions that affect almost the entire business (marketing, production process, etc.).

It improves efficiency and productivity. This is probably what concerns most of you—and what you are under the most pressure to achieve. Think about this. You have technological tools to make your work easier, and you have more information, which allows you to make better increase in productivity decisions. When used intelligently, the digitalization of business can lead to a significant and can reduce costs.

In summary, *digitalization creates an environment for digital business, whereby digital information is at the core.* Those who deprioritize or ignore digitalization do so at their own peril, especially when the risks of having nimbler, untethered disrupters surpass them are all too real.

Digital Transformation

If business leaders think they can digitize a business or digitalize enough processes to digitally transform themselves, they are misunderstanding the terms and missing out on opportunities to evolve, gain competitive advantage, respond to consumer and employee expectations and demands, and become agile businesses.

However, the reality is that few businesses have undergone successful digital transformations. Gerald Kane and his colleagues at Boston College's Carroll School of Management, in their 2017 global study of digital transformation, (2) found that only 25% of organizations had transformed into digital businesses, 41% were on transformative journeys, and 34% invested more time talking about the trend than they did acting on it. It is noteworthy that 85% of executives stated that attaining digital maturity is critical to organizational success. The discrepancy between recognizing digitalization as a competitive necessity and successfully implementing a transformative strategy suggests that many leaders are unsure how to harness the opportunities that a digital transformation brings to people, processes, and technology. In this report, Kane and his colleagues show a lot of room for improvement.

> Those who deprioritize or ignore digitalization do so at their own peril, especially when the risks of having nimbler, untethered disrupters surpass them are all too real.

Why do Organizations Hesitate?

So why do companies balk at adopting digital transformation? Is it the "people" element? In our experience, this is always the most difficult, yet it often gets delegated to the HR department to deal with.

Is it because achieving full buy-in to the new business model requires changing the organization's culture? Until digital transformation practice is adopted at the department level, in other words *internalized,* the transformation is not complete. For companies undergoing digital transformation, high levels of internalization—that is, organization-wide commitment to a practice—are shown to complement and strengthen the relationship between a practice's implementation and its success. (3)

Or is the obstacle the deceptively hard job of innovating and putting customers' needs at the center of these new business models? Seems simple, doesn't it? Put technology aside and focus only on the new needs of customers—particularly new customers. Think of Amazon and Tesla at this point. These organizations are the benchmark of disruptive digital businesses that are transforming themselves, seemingly continuously.

Whatever the reasons, these actions represent substantial changes and *risk* to incumbent organizations. For this and other reasons, the topics of digital transformation and digital leadership have been the mainstay of almost every consulting publication and management journal. This shows a profound interest, if not an outright economic need, to better define, understand, and manage this phenomenon. Yet despite the promise of creating new and productive business designs, designs that leverage the explosion occurring in communications and computing technologies, many organizations exhibit a wait-and-see attitude to digital transformation. While a few companies have achieved front-runner status, the majority lag behind.

> Some senior managers actually worry that their organizations may not have the knowledge, or even the will, to undergo a change of this magnitude.

At best, the digital transformation activities of these organizations can be better labeled experiments or proofs of concept. Some senior managers actually worry that their organizations may not have the knowledge, or *even the will*, to undergo a change of this magnitude.

Unfortunately, risk aversion still seems to rule the day.

Fears of Senior Management

The Enterprise Risk Management Initiative of the North Carolina State University Poole College of Management collaborates with Protiviti Consulting each year to assess the top risks facing organizations. In their recent 2019 survey (NC State ERM Initiative and Protiviti, 2019), they interviewed 825 members of the top management teams—the "C" officers and board members alike—representing industries from around the world. In the survey, 68% of respondents rated "rapid speed of disruptive innovation" as the top strategic threat to their organizations. (4) These organizations are concerned that new technologies may emerge that will outpace their ability to keep up or remain competitive. With the advent of new digital technologies and rapidly changing business models, these companies worry that their present organizations are not agile enough to respond to new customer expectations that may change their core business model.

For example, Blockbuster once controlled the majority share of the movie rental business but failed to adapt their business model to account for digital platforms (read Netflix), which resulted in Blockbuster declaring bankruptcy in 2010. History has taught us that as business model disruptors emerge, companies are justifiably concerned that their organizations will not make the timely changes needed to remain competitive.

Tightly linked to the emergence of disruptive innovations, respondents also identified their organizations' overall "resistance to change" as their top operational risk. Respondents are becoming increasingly concerned with their organizations' lack of willingness to change the business model and alter core operations in response to changes in the business environment or industry. In recent years, companies have learned that mistakes in the digital economy can be lethal. If major business model disruptors emerge, these companies are concerned that their organizations will not make the timely changes to remain competitive.

"Our culture may not encourage the timely escalation of risk issues." Culture has been a chief concern of industry leaders since 2015 and has been a top 10 enterprise risk ever since.

> Culture has been a chief concern of industry leaders since 2015 and has been a top 10 enterprise risk ever since.

Combined with resistance to change, a poor culture can be equally lethal for organization leaders if they fall out of touch with business realities.

In 2020, the issue was heightened when ERM and Protiviti cited two major, interlocking themes that surround seven of the top ten risks facing corporations: "Talent and culture" and "technology and innovation." (5) These two themes are especially relevant to companies trying to advance their digital maturity and capture the transformative potential of technology.

Talent and Culture

Two of the top ten risks deal with the ability of these firms to attract and retain top technical talent. A third deals with concerns associated with the succession of top technical talents in tightening markets. Risk number 10 dealt specifically with the reskilling, or *upskilling,* of people knowledgeable in AI-enabled digital technologies that will change the future of work. Lastly, two other risks present implications to the organization's culture—"resistance to change" and "the culture surrounding the escalation of risks concerns."

Technology and Innovation

Three of the top ten risks have implications about the organizations' ability to embrace and manage technology and innovation. For example, "existing operations, legacy IT systems, and digital capabilities might not be sufficiently flexible to compete." Furthermore, respondents are concerned that technology impacting customers' allegiance to brands is "failing to keep pace with market developments."

In their book *Digital Leadership,* (6) Creusen, Gall, and Hackl emphasize that the competence factor will become increasingly important in the context of digital transformation: "A Chief Digital Officer (CDO) is important, not because of the title on the business card, but because of the competence to implement the digital transformation in a company." Gerald Kane points out that the creation of a CDO position signals the strategic nature of digital transformation for the entire organization. He urges organizations to hire digital leaders to "get the ball rolling." Many organizations, he argues, have overlooked digital transformation for so long that they don't even know where to begin. But they can make so-called anchor hires to catalyze the process. These are outside leaders with deep digital transformation experience who can provide the needed expertise and perspective. (7)

Bringing these skills in from the outside will support the organization's transformation, but it is also important to sustain what has been built. For that, Kane says, you'll need an entirely different operational skill set, which you may well find internally.

In summary, talent, culture, technology, and innovation blanket corporate risks. These themes are inextricably connected. Companies that unable to attract, train, and retain the *right* digital leaders are less likely to execute increasingly complex strategies for navigating new digital technology-based business environments.

Chapter References

1. Nambisan, S., Iyytinen, K., Majchrzak, A., & Song, M. (2017). Digital Innovation Management: Reinventing Innovation Management Research in a Digital World. *MIS Quarterly*: 41.
2. Kane, G.C., Palmer, D., Phillips, A.N., Kiron, D., & Buckley, N. (2017). Achieving Digital Maturity: Adapting Your Company to a Changing World. *Deloitte Insights*.
3. Abbu, H. R. & Gopalakrishna, P. (2019). Synergistic effects of market orientation implementation and internalization on firm performance. *Journal of Business Research*.
4. NC State ERM Initiative and Protiviti. (2019). *Executive Perspectives on Top Risks for 2019: Key Issues Being Discussed in the Boardroom and C-Suite*.
5. NC State ERM Initiative and Protiviti. (2020). *Executive Perspectives on Top Risks for 2020: Key Issues Being Discussed in the Boardroom and C-Suite*.
6. Creusen, U., Gall, A.B. & Hackl, O. (2017) *Digital Leadership: Führung in Zeiten des digitalen Wandels*. Wiesbaden: Springer Gabler.
7. Kane, G.C., Phillips, A.N., Copulsky, J. & Andrus, G. (2019). How Digital Leadership Is(n't) Different. *MIT Sloan Management Review*, 60(3), (Spring 2019): 34-39.

CHAPTER 2

SCIENCE OF TRUST

"Trust is not a benefit that comes packaged with the nameplate on your door. It must be earned, and it takes time. As a leader, you are trusted only to the degree that people believe in your ability, consistency, integrity, and commitment to deliver." (1)

– David Horsager

"Trust is mandatory for optimization of a system. Without trust, each component will protect its own immediate interests to its own long-term detriment, and to the detriment of the entire system." (2)

– W. Edwards Deming

We all recognize the importance of trust in any relationship. When it comes to organizational relationships, a culture of trust has been shown to yield higher engagement, happier employees, greater productivity, and higher profits. More on this later.

In this chapter, we examine the science of trust through the lens of scholarly research conducted during the last six decades. We are not intending to embark on an in-depth scientific literature review. Instead, the primary aim of this chapter is to present the prominent thoughts of these opinion shapers and provide us with their learned perspectives on the nature—and functions—of trust within organizations. We believe this provides the reader with a high-level overview of the science behind conceptualizing trust, its antecedents, and its consequences in organizations.

Implications for Management

Companies need digital leaders if they are to succeed. Furthermore, these leaders must be able to deal with the people issues. They must recognize and be ready to change the organization's culture. This is a tall order. A 2018 McKinsey survey revealed, "Culture and the associated behavioral changes were assessed as the main obstacle to digital effectiveness." They stressed that "executives who wait for organizational cultures to change organically will move too slowly as digital penetration grows, blurs the boundaries between sectors, and boosts competitive intensity." (3)

A strong top-down direction from the senior executive team coupled with methods that engage employees in making the change happen is the only effective way to drive digital transformation. In fact, Satya Nadella, CEO of Microsoft, points out the impact of culture on digital transformation: "Culture change is not an abstraction; it is really walking the walk." (4) Senior management must step up to cultural impediments and overcome their top operational challenge—the organization's resistance to change.

We know that leaders of Digitally Mature companies, compared to leaders of Digitally Developing companies, collaborate with cross-functional counterparts to achieve the level of trust that is needed to go through a digital transformation. According to Teichmann and Hüning, "Digital Leadership stands for everything that a lot of organizations currently lack—innovative spirit, value orientation, potential for disruption and contradiction, flexibility in the matter at hand, but also steadfastness in essence, a high level of social competence and a great deal of courage." (5)

Defining Trust

The concept of trust has been explored across many disciplines by a number of scholars dating back several decades. We find several definitions of trust in literature, from varied settings and perspectives. Let's focus on a few here.

Mayer, Davis, and Schoorman (6) defined trust as "the willingness to be vulnerable to another party when that party cannot be controlled or monitored." Decades prior, Deutsch (7) put forth a thesis that risk, or having something invested, is a prerequisite to trust. Along the same lines, Rousseau and colleagues (8) defined trust as "a psychological state comprising the intention to accept vulnerability based upon positive expectations of the intentions or behaviors of another."

We selected these definitions because they recognize the relationship between trust and the critical issue of taking risk. More trust leads to more risk-taking behaviors on the part of the trustor. Risk is inherent in vulnerability, and the trustor's behaviors allow vulnerability to the trustee, rather than willingness to be vulnerable alone. Trust, therefore, is a generalized behavioral intention to take risk, whereas its outcome is actually taking risk itself!

Factors that Promote Trust: Antecedents

Researchers for decades have explored the factors that enable trust within organizations. They have put leaders' *character* and *competency* at front and center.

Mayer and colleagues highlight three essential characteristics of trustworthiness in an integrative model of organizational trust (Figure 2).

1. *Ability*: perception that a trustee has skills and competencies in the domain of interest
2. *Benevolence*: trustor's (i.e., the trusting party's) perception that the trustee cares about the trustor, and
3. *Integrity*: perception that the trustee adheres to a set of principles that the trustor finds acceptable.

ABILITY

Perception that a trustee has
skills and competencies in
the domain of interest

↗ ↖

BENEVOLENCE

Trustor's perception that
the trustee cares about
the trustor

↔

INTEGRITY

Perception that the
trustee adheres to a set of
principles that the trustor
finds acceptable

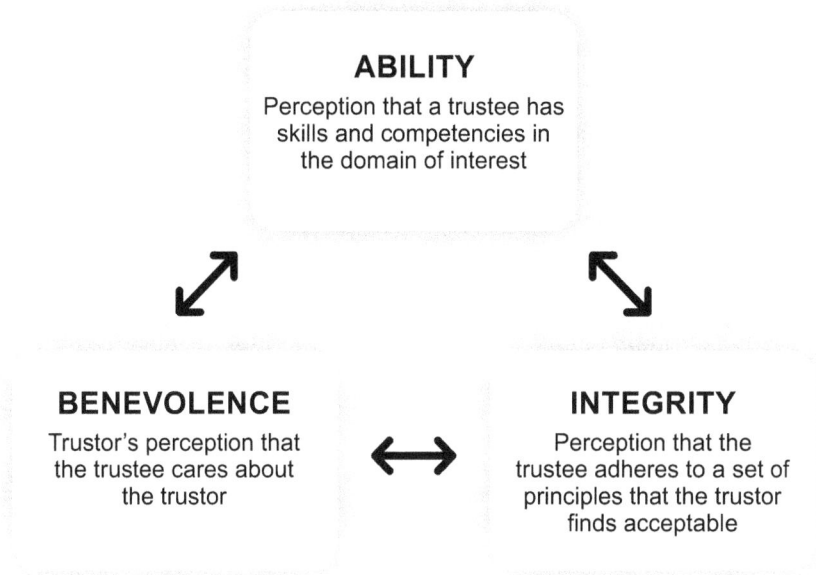

Figure 2. *Integrative Model of Organizational Trust*

As a set, these three appear to explain a major portion of trustworthiness. Each contributes a unique perspective from which to consider the leader (trustee), while the set provides a solid understanding of the antecedents of trust.

Similarly, Gabarro (9) identified nine "bases" of trust through a series of clinical interviews. They include:

1. *Integrity (honesty and moral character)*
2. *Motives (intentions and agenda)*
3. *Consistency of behavior (reliability)*
4. *Openness (leveling and expressing ideas freely)*
5. *Discreetness (keeping confidences)*
6. *Functional/specific competence (knowledge and skills related to a specific task)*
7. *Interpersonal competence (people skills)*

8. *Business sense (common sense and wisdom about how a business works), and*
9. *Judgment (ability to make good decisions)*

Butler (10) used Gabarro's findings along with other studies of managerial trust and manager interviews to develop a content theory of trust conditions and to derive scales to measure them.

David Horsager (1), author of *The Trust Edge: How Top Leaders Gain Faster Results, Deeper Relationships, and a Stronger Bottom Line,* highlighted eight key strengths that earn trust over time.

1. *Clarity: Be clear about your mission, purpose, expectations, and daily activities*
2. *Compassion: People are often skeptical about whether someone really has their best interests in mind*
3. *Character*
4. *Contribution (results)*
5. *Competency (stay fresh, relevant, and capable)*
6. *Connection (genuine relationships)*
7. *Commitment, and*
8. *Consistency*

Several other opinion shapers have emphasized similar factors to build trust in organizations.

- Hovland, Janis, and Kelley (11), in the famous Yale studies, emphasized communication and attitude change as key characteristics of a trustworthy leader. According to these researchers, expertise and trustworthiness resulted in credibility of a leader.
- Deutsch (12) argued that the "individual must have confidence that the other individual has the ability and intention to produce it."
- Rotter (13) highlighted that interpersonal trust "is an expectancy held by an individual or a group that the word, promise, verbal or written statement of another individual or group can be relied upon."

- Gabarro (9), stressed predictability: "the extent to which one person can expect predictability in the other's behavior in terms of what is 'normally' expected of a person acting in good faith."
- Cook and Wall (14) defined trust as "the extent to which one is willing to ascribe good intentions to and have confidence in the words and actions of other people."
- Butler (10) included consistency, integrity, and fairness as conditions of trust.

Through the teachings of these various opinion shapers, we see two foundations emerge for leaders to build trust: *competence* and *character*.

A Pattern Emerges

The first foundation is *competence*. It enables a party to have influence within some specific domain. This involves perceptions that the other party has the knowledge and skills needed to do a job, along with the interpersonal skills needed to succeed.

In the context of developing knowledge workers, it is important that leaders pay additional attention to knowledge-building behaviors, such as scanning the environment for new ideas, developing knowledge networks, sharing technical expertise, bringing in outside experts in areas where you lack experience, providing feedback that is relevant to increasingly complex tasks, and overseeing the quality of work that you may have not done yourself. Together, demonstrating competence in these skills engenders trust and knowledge sharing, and these competence-enhancing behaviors play an important role in building trust. (15)

The second foundation is *character*, which includes two distinct constructs: benevolence and integrity. Benevolence is the extent to which a trustee is believed to want to do good for the trustor. It is the perception of a positive orientation of the trustee toward the trustor. Integrity is the degree to which a trustee is believed to follow sound ethical principles. Leaders must develop perceptions of benevolence through coaching behaviors that foster a supportive context. Finally, leaders must develop and sustain perceptions of integrity by acting in ways that are consistent with their values and accountability.

Consequences of Trust

Paul J. Zak, (16) Harvard researcher and author of *The Trust Factor: The Science of Creating High Performing Companies*, discovered that "compared with people at low-trust companies, people at high-trust companies report 74% less stress, 106% more energy at work, 50% higher productivity, 13% fewer sick days, 76% more engagement, 29% more satisfaction with their lives, and 40% less burnout." Zak also showed, through a series of "neuroscience of trust" experiments, that there is a direct correlation between the amount of oxytocin a person's brain produces and the level of trust they feel in any given situation.

Chris Argyris (17), an American business theorist, proposed back in 1964 that trust in management is important for organizational performance. Organizational trust can harmonize the employment relationship, employees' job satisfaction, and organizational commitment, and companies with higher level of employee trust will face comparatively less resistance when implementing new organizational initiatives. (18)

Within organizational settings, trust can function to reduce transaction costs. In addition, it can increase spontaneous sociability among organizational members and facilitate appropriate forms of deference to organizational authorities. When the employee has trust in the top manager, his organizational citizenship behaviors may benefit the whole organization. (19)

Trust in various levels of management affects an employee's ability to focus attention on activities that add value to an organization. In other words, when employees lack trust in management, they are unwilling to be vulnerable to management and are preoccupied with nonproductive issues, including self-protection and defensive behaviors.

Trust and Digital Leadership

Recently, the concept of trust has been employed in the context of digital transformation and digital leadership as well. Given the digital realm, the competence factor has become increasingly important, especially in the context of digital transformation. For example, the regular achievement of results strengthens the competence aspect. The track record of a leader helps build trust. "Competence is a key factor when hiring a Chief Digital Officer (CDO) when it is necessary to deploy digital transformation in a company." (20) Furthermore, the creation of a CDO position signals the strategic nature of digital transformation for the entire organization. (21)

> When employees lack trust in management, they are unwilling to be vulnerable to management, and are preoccupied with nonproductive issues, including self-protection and defensive behaviors.

Kane and colleagues, (22) in their long-term research on digital transformation with over 20,000 business executives, managers, and analysts around the world, found four main focal points of digital leadership.

1. *Transformative Vision*: This includes knowledge of markets and trends, business acumen, and problem-solving skills.
2. *Forward-Looking Perspectives*: This is specified as clear vision, sound strategy, and foresight.
3. *Digital Literacy*: This means the pre-existing experience and knowledge about digital technologies such as data and analytics, Artificial Intelligence (AI), blockchain, etc. Digital literacy helps C-Suite managers to anticipate crucial emerging trends.

4. *Adaptability*: This helps leaders respond to a fluid environment and change course if the technology and market environments evolve in unanticipated ways. This mindset also enables a digital leader to continually update his or her knowledge stores to account for changes in technology and avoid obsolescence.

Digital leadership is a fast, cross-hierarchical, team-oriented, and cooperative approach, with a strong focus on innovation. The personal competence of the leader, including their mindset and their ability to apply new methods and instruments, are critical dimensions for digital leaders (23).

Covey's Trust Model

We take inspiration from Stephen M.R. Covey's Trust Model (see Figure 3) to conceptualize digital leadership. We know that trust enables companies to succeed in their communications, interactions, and decisions, and to move with incredible speed. True transformation starts with building *credibility* at the personal level, which is the foundation of trust. When a leader's credibility and reputation are high, it enables them to establish trust fast. As a result, speed goes up, and cost goes down. (24)

> When a leader's credibility and reputation are high, it enables them to establish trust fast. As a result, speed goes up, and cost goes down.

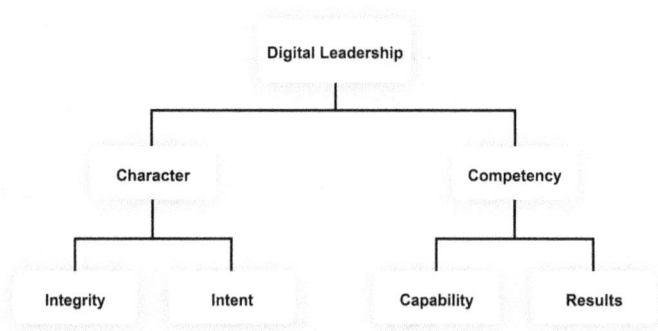

Figure 3. *Four Cores of Digital Leadership (Adapted from Covey's Trust Model)*

Stephen M.R. Covey breaks *character* down into different character traits. He emphasizes two factors: integrity and clarity of the leader's intentions. Integrity is walking the talk. It also includes honesty, humility, courage, and congruence. Clarity of the leader's intention means the absence of hidden agendas. The *competencies* category emphasizes the individual capabilities of the leader, as well as the regularity with which results are achieved.

Covey's Trust Model is elegant but a bit complex. The system, as we would prefer to call it, contains a set of "four cores of credibility" as well a process for how these values fan out across "five waves of progression"—from self-trust to societal trust. As part of this progression, Covey also identifies "thirteen behaviors" that are needed to embed and reinforce these values at every level of wave.

Four Cores of Credibility

From the literature we determined that trust has two major dimensions—*character* and *competence.*

1. *Character is a constant.* It is necessary for trust under any circumstances.
2. *Competence is situational.* It depends on what the circumstances require.

Covey takes this concept further and defines four cores of credibility: *Integrity, intent, capabilities,* and *results* that underpin these dimensions. *Integrity* and *intent* are *character* cores. *Capabilities* and *results* are *competency* cores. He asserts that all four cores must work in tandem and are key to building credibility, i.e., being trustworthy (Figure 3).

The First Core Is Integrity

Covey asserts that *integrity* is more than honesty. In addition to honesty, integrity is made up of three other virtues: *Congruence, humility,* and *courage.*

29

Congruence is when one acts according to his values, when there is no gap between what one intends to do and what one actually does. Humility is the ability to look out for the good of others in addition to what is good for you. And courage is the ability to do the right thing even when it may be difficult. It is when you do what you know is right regardless of the possible consequences.

The Second Core Is Intent

Intent springs from our *character*. It is how we know we should act. Covey breaks *intent* down to three things: *Motive, agenda,* and *behavior.*

Motive is why you do what you do. The best motive in building trust is genuinely caring about people. Agenda stems from our motive. The best agenda is honestly seeking what is good for others. Behavior is putting your agenda into practice. It is what we do based upon what we intend to do and what we are actively seeking. People see and judge leaders through their behavior.

The Third Core Is Capabilities

Capabilities inspire the trust of others, particularly when they are specifically those needed for the task at hand. Our capabilities also give us the self-confidence that we can do what needs to be done. Covey breaks *capabilities* down to five things: *Talents, attitude, skills, knowledge, and style.*

Talents are the things we naturally do well. Attitude is how you see things. Skills are the things you have learned to do well. Knowledge is what you know and continue to learn. Style is your unique way of doing things.

The Fourth Core is Results

Results are the deliverables people look at to judge your credibility. They are what you contribute to the organization. You can't hide from your results. Lack of results would mean a lack of credibility and trust. Covey breaks *results* down to three things: *past results, current results,* and *potential results.*

First, your past results: what you have proven you can do. Second, your current results: what you are contributing right now. Third, your potential results: what people anticipate you will accomplish in the future.

> Results are the deliverables people look at to judge your credibility. They are what you contribute to the organization. You can't hide from your results. Lack of results would mean lack of credibility and trust.

In considering results, leaders need to ask two critical questions:

1. What results are we getting?
2. How are we getting those results?

To increase trust, leaders must effectively communicate results so that people become aware of them.

Five Waves of Trust

Trust is built from the inside out. Whatever trust we are able to create in our organizations or in the marketplace is a result of the credibility we first create in ourselves, in our relationships, in our organizations, in our markets, and in society. (24)

In his Five Waves of Trust, Covey identifies the 13 behaviors, in his experience, needed to embed and reinforce the core values at every level—self, relationship, organizational, market and societal trust.

1. Self-Trust

The first wave, self-trust, deals with the confidence we have in ourselves, in our ability to set goals, to keep commitments, to walk our talk, and also in our ability to inspire trust in others. The whole idea is to become, both to ourselves and to others, a person who is worthy of trust.

The key principle underlying this wave is credibility, which comes from the Latin root *credere*, meaning "to believe." In this first wave, the Four Cores of Credibility can be used to increase credibility and to firmly establish trust with ourselves and with others. The end results of high character and high competence is credibility, judgment, and influence.

2. Relationship Trust

The second wave, relationship trust, is about how to establish and increase the "trust accounts" we have with others. The key principle underlying this wave is consistent behavior, and there are 13 Behaviors of High Trust that are based on principles that govern trust in relationships. We don't want to discuss the 13 behaviors at this point because we don't want to interrupt the flow of the 5 waves, and they apply equally to other waves of trust. (Please see Thirteen Behaviors of High Trust below.)

3. Organizational Trust

The third wave, organizational trust, centers around alignment. Most people find that their organization has symptoms of low trust—people manipulating facts, withholding information, resisting new ideas, and covering up mistakes. The low trust environment is a result of violating these principles, both individually and organizationally.

Leaders are missing the solution because they are not looking at the systems, structures, processes, and policies that affect day-to-day behaviors. They are focused on the symptoms instead of the principles that promote trust.

4. Market Trust

The fourth wave, market trust, is all about brand or reputation. It's all about the feeling you have that makes you want to buy products or services or invest your money or time. This is the level where most people clearly see the relationship between trust, speed, and cost.

Brand is important to all organizational entities, including governments, school districts, charities, and hospitals, not to mention corporations. If your organization does not have the brand it desires, leaders can measurably increase its value by using the Four Cores as a diagnostic tool to pinpoint the reason why, and where investment will bring the greatest return. Then use the 13 Behaviors with external stakeholders—customers, suppliers, investors, communities—and build trust at the marketplace level.

5. Societal Trust

The fifth wave, societal trust, is based on the overriding principle of contribution. It's intent is to create value instead of destroying it.

This is not an impractical or utopian view of the world. The principles of contribution and responsibility create trust at a societal level through today's trend toward global citizenship or corporate social responsibility.

Though initially, fear might motivate global citizenship, over time, the dividends and abundance created by contribution will become primary drivers for both individuals and organizations. Global citizenship will be demanded as customers support companies that demonstrate the Four Cores.

Thirteen Behaviors of High Trust

Covey (24) lists 13 behaviors that can go a long way in building an environment of trust. The first five flow initially from character, the second five from competence, and the last three flow from both. They are practitioner-based and validated by research. These 13 behaviors can be learned and applied by any individual at any level within any organization. The net result is a significantly increased ability to generate trust with all involved to enhance relationships and achieve better results.

Behavior #1: Talk Straight

Talking straight goes a long way to inspire trust. We often see counterfeit or fake behavior in managers who beat around the bush, withhold information, double-talk, flatter individuals, or "technically" tell the truth. You probably know people yourself who frequently tell "half-truths".

Behavior #2: Demonstrate Respect

Leaders must show respect, fairness, and civility to individuals. The counterfeit response is to fake respect or to show respect for some (those who can do something for us), but not for all (those who can't). We're sure you have seen this picture before.

Behavior #3: Create Transparency

Leaders must be transparent. They must be open and genuine and tell the truth in a way people can verify. It's based on honesty and openness. The opposite of this behavior is to hide, cover, or obscure. It includes hidden agendas, hidden meanings, and often hidden objectives. These types of behaviors can be disastrous to digital transformation.

Behavior #4: Right Wrongs

This is more than simply apologizing. Leaders must take the first step to making up for and doing what they can to correct the mistake. This behavior is based on humility and integrity. Its opposite is to deny or justify wrongs or fail to admit mistakes until they are forced to do so.

Behavior #5: Show Loyalty

Leaders must give credit to others and acknowledge them for their part in bringing about results. They should not appear to give credit to someone when they're present yet downplay their contribution and take all the credit when they're not there. Another dimension of showing loyalty is to speak about others as if they *are* present. The counterfeit behavior is indulging in sweet-talk in front of people and bad-mouthing them behind their backs.

Behavior #6: Deliver Results

This behavior grows out of the principles of responsibility, accountability, and performance. The counterfeit is engaging in "activities" instead of delivering results. Results are always judged in relation to expectations. It is important in each situation that leaders define the results that will build trust, and then deliver those results—consistently, on time, and within budget. Note, leaders must make sure the organization thoroughly understands their expectations. If leaders really want to build trust, employees have to know what "results" mean to the customer. To over-promise and under-deliver often leads to the depletion of trust.

Behavior #7: Get Better

This behavior is based on continuous improvement, learning, and change. In seeking to "get better", there are two strategies that are particularly helpful: seek feedback and learn from mistakes. The next time the team makes a mistake, leaders must identify the learning from it and the ways they can improve the approach to get different results next time. Leaders must also encourage others to take appropriate risks and to learn from failure.

Behavior #8: Confront Reality

This behavior is about taking the tough issues head-on, sharing the bad news as well as the good. When leaders openly confront reality, they facilitate open interaction and fast achievement. Instead of having to wrestle with all the hard issues on their own—while trying to paint a rosy picture for everyone else—they actually engage the creativity, capability, and synergy of others in solving those issues.

Behavior #9: Clarify Expectations

Leaders must *render explicit* the goals and objectives of the transformation strategy. The opposite of this behavior is to leave expectations undefined, to assume they're already known or to fail to disclose them so there is no shared vision of the desired outcomes. When results are delivered but not valued, everyone is disappointed and trust declines. In every interaction there are expectations. And the degree to which these expectations are met or violated affects trust. In fact, unclarified expectations are often the cause of broken trust.

Behavior #10: Practice Accountability

Leaders must look for ways to create an environment of accountability. Trust results when people know that everyone will be held to certain standards. When leaders don't hold people accountable, it creates a sense of disappointment, inequity, and insecurity. Leaders must practice accountability by holding their direct reports accountable for their actions. Again, leaders must clarify their expectations first so that everyone knows what they're accountable for and by when.

Behavior #11: Listen First

Leaders must learn to listen; and to understand first. Otherwise, they may be acting on assumptions that are totally incorrect—acting in ways that turn out to be embarrassing and counterproductive. Listening is not just hearing what is said. So, *listen first* means to listen not only with the ears but also with the eyes and the heart. This may be one of the toughest—but most important—behaviors for digital leaders to master.

Behavior #12 Keep Commitments

This is the quickest way to build trust in any relationship. To break commitments or violate promises is the quickest way to destroy trust. Keeping commitments is the perfect balance of character and competence. Particularly, it involves integrity (character) and the ability to do what we say we are going to do (competence).

Behavior #13 Extend Trust

This behavior is based on empowerment, reciprocity, and a fundamental belief that most people are capable of being trusted, want to be trusted, and will do well when trust is extended to them. Leaders must do two things to avoid a complete breakdown in trust. Don't extend "false trust," i.e., give people the responsibility but not the authority or resources to get a task done. Good managers don't act like they trust someone when they really don't.

Taxes and Dividends of Trust

In every relationship, what we do has far greater impact than anything we can say. Good words signal behavior, declare intent, and can create enormous hope. And when those words are followed by appropriate behavior, they increase trust, sometimes dramatically.

For businesses, today's global marketplace puts a premium on true collaboration and all these interdependencies require trust. Partnerships based on trust outperform partnerships based on contracts. *Compliance does not foster innovation, trust does!*

Covey makes a compelling argument that when trust is low, in a company or in a relationship, it places a hidden "tax" on every transaction: every communication, every interaction, every strategy, every decision is taxed, bringing speed down and sending costs up. Covey identifies seven taxes: redundancy, bureaucracy, politics, disengagement, turnover, churn, and fraud in low-trust organizations.

By contrast, individuals and organizations that have earned and operate with high trust experience the opposite of a tax—a "dividend" that is like a performance multiplier, enabling them to succeed in their communications, interactions, and decisions and to move with incredible speed. Covey also identifies the seven dividends: increased value, accelerated growth, enhanced innovation, improved collaboration, stronger partnering, better execution, and heightened loyalty in high-trust organizations.

Trust is the one thing that affects everything else you're doing. It's a performance multiplier that takes your trajectory upwards for every activity you engage in, from strategy to execution. The distrust we see all around is suspicion, a response to the corporate scandals and vicious downward cycles of cynicism. Executives need to understand the economic benefits of this *trust dividend*, especially when the behavior is real, not artificially or superficially created as PR to manipulate trust. Leaders must lead in creating trust, and the job of the leader is to go first. Someone needs to go first, and that's what leaders do. Leaders go first.

> Trust is the one thing that affects everything else you're doing. It's a performance multiplier that takes your trajectory upwards for every activity you engage in, from strategy to execution.

The Main Takeaway – The Ability of Leaders Determines Success

Before going further, let's look at the extended version of Covey's Trust Model. In Chapter Four, we use this framework as the basis for our interview guide. See Figure 4, Digital Leadership Extended Model.

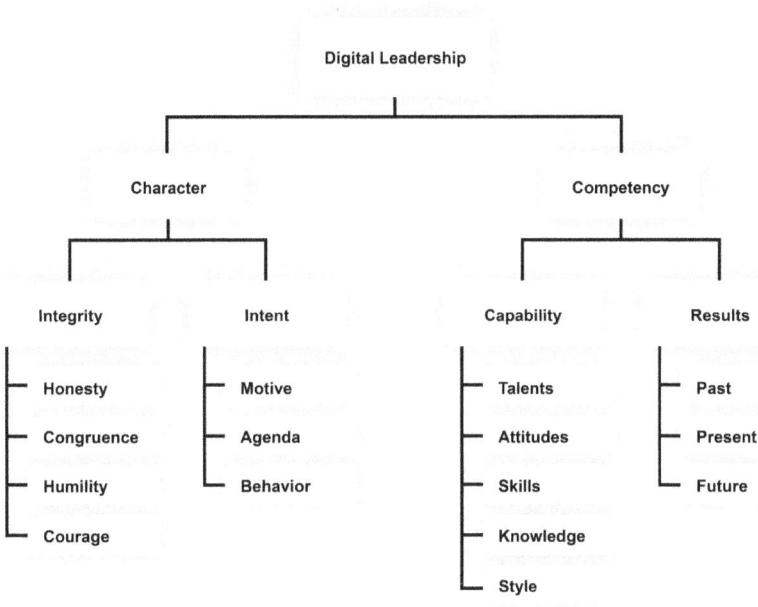

Figure 4. *Digital Leadership Extended Model (Adapted from Covey's Trust Model)*

Using this model, we interview digital leaders who have already led digital transformations. We hope they will help us understand how they apply these factors to achieve their organizations' transformation.

In our view, the difference between the organizations that succeed at digital transformation and those that do not—or cannot—goes right back to the leaders of these organizations, and particularly their ability to build and maintain the trust of their employees through these often perilous journeys.

The enduring human traits of these leaders far outweigh their proficiency in the always-evolving field of information technology. We further believe the ability of these leaders to engender the trust of their employees allows them to accomplish a digital transformation much faster and at much less cost to the organization. We all agree these metrics are paramount in today's economy. The rest of the book is devoted to demonstrating this premise. The book is focused on the significant and positive economic and cultural effect digital transformation can have on organizations of all types—*when it has the right leadership*.

ERM and Protiviti researchers concluded that the question is not *if* digital is going to up-end their current business model, but rather *when*. Even when executives are aware of emerging technologies that obviously have disruptive potential, it is often difficult to have the vision or foresight to anticipate the nature and extent of change. Overcoming these hurdles represents the major challenge leadership teams confront when digitizing their businesses. However, there is no other choice because the trend towards digital transformation will not weaken. On the contrary, according to Westerman and his collaborators, "The innovations and disruptions of the past ten years have been nothing short of astonishing… they're just the warm-up acts for what's to come." (25)

In this chapter, we took you on a journey to explore science of trust—its origins, themes, opinion shapers, antecedents, and consequences. Whether it is the 1958 definition from Deutsch or Covey's four cores of credibility from his 2006 masterpiece, *The Speed of Trust*, and everything in between, the key theme is clear. *The competency and character of the leader is critical to build trust within organizations.*

Chapter References

1. Horsager, D. (2012). *The Trust Edge: How top leaders gain faster results, deeper relationships, and a stronger bottom line.* Minneapolis, MN: Summerside Press.
2. Deming, W. E. (1994). *The trust factor: Liberating profits and restoring corporate vitality.* New York: McGraw-Hill.
3. McKinsey Survey (2018). *Unlocking Success in Digital Transformations.*
4. Nadella, S., & Euchner, J. (2018). Navigating digital transformation. Conversations. *Research-Technology Management* 61(4).
5. Teichmann, S. & Hüning, C. (2018). Digital Leadership – Führung neu gedacht: Was bleibt, was geht?' in Keuper, F. et al. (eds.) *Disruption und Transformation Management: Digital Leadership – Digitales Mindset – Digitale Strategie.* Wiesbaden: Springer Fachmedien Wiesbaden, pp. 23–42.
6. Mayer, R. C., Davis, J. H., & Schoorman, F. D. (1995). An integrative model of organizational trust. *Academy of Management Review*, 20: 709 –734.

7. Deutsch, M. (1958). Trust and suspicion. *Journal of Conflict Resolution*, 2: 265–279.

8. Rousseau, D. M., Sitkin, S. B., Burt, R. S., & Camerer, C. (1998). Not so different after all: A cross-discipline view of trust. *Academy of Management Review*, 23, 393–404.

9. Gabarro, J.J. (1978). The development of trust and expectations. In *Interpersonal Behavior: Communication and Understanding in Relationships*, Athos, A.G. & Gabarro, J.J. (eds.), pp. 290–303. Englewood Cliffs, NJ: Prentice Hall.

10. Butler, John K., Jr. (1991). Toward Understanding and Measuring Conditions of Trust: Evolution of a Conditions of Trust Inventory. *Journal of Management*, 17, 3.

11. Hovland, C. I., Janis, I. L., & Kelley, H. H. (1953). *Communication and persuasion; psychological studies of opinion change.* Yale University Press.

12. Deutsch, M. (1960). The Effect of Motivational Orientation upon Trust and Suspicion. *Human Relations*, 13(2), 123–139.

13. Rotter, J. B. (1967). A new scale for the measurement of interpersonal trust. *Journal of Personality*, 35: 651-665.

14. Cook, J., & Wall, T. (1980). New work attitude measures of trust, organizational commitment and personal need non-fulfillment. *Journal of Occupational Psychology*, 53, 39–52.

15. Bligh M.C. (2017). Leadership and Trust, in Marques J. & Dhiman S. (eds.) *Leadership Today*. Springer Texts in Business and Economics.

16. Zak, P.J. (2017). The neuroscience of trust. *Harvard Business Review,* 95(1): 84–90.

17. Argyris, C. (1964). *Integrating the individual and the organization*. New York: Wiley.

18. Dirks, K. T., & Ferrin, D. L. (2002). Trust in leadership: Meta-analytic findings and implications for research and practice. *Journal of Applied Psychology*, 87(4), 611–628.

19. Kramer, R. M. (1999). Trust and distrust in organizations: Emerging perspectives, enduring questions. *Annual Review of Psychology*, 50, 569–598.

20. Creusen, U., Gall, A. B., & Hackl, O. (2017). *Digital Leadership: Führung in Zeiten des digitalen Wandels*. Wiesbaden: Springer Gabler.

21. Vial, G. (2019). Understanding digital transformation: A review and a research agenda. *The Journal of Strategic Information Systems*, 28(2), 118–144.

22. Kane, G.C., Phillips, A.N., Copulsky, J.R., & Andrus, G. R. (2019). *The Technology Fallacy: How People Are the Real Key to Digital Transformation*: MIT Press.

23. Oberer, B., & Erkollar, A. (2018*)*. Leadership 4.0: Digital Leaders in the Age of Industry 4.0. *International Journal of Organizational Leadership*, 7(4), 404–412.

24. Covey, S. M. R. (2006). *The Speed of Trust: The one thing that changes everything*. Simon & Schuster.

25. Westerman, G., Bonnet, D., & McAfee, A. (2014). *Leading Digital: Turning technology into business transformation*. Boston, Massachusetts: Harvard Business Review Press.

CHAPTER 3

DIGITAL MATURITY DEMANDS DIGITAL LEADERS

"Having a personality of caring about people is important. You can't be a good leader unless you generally like people. That is how you bring out the best in them."

– Sir Richard Branson

Digital Leaders—How and Why are They Different?

What exactly is digital leadership? More importantly, how do digital leaders differ from traditional leaders? To answer these questions, we must first understand, what does digital leadership solve?

What the Experts Say

Digital leadership stands for everything that a lot of organizations currently lack: "Innovative spirit, value orientation, potential for disruption and contradiction, flexibility in the matter at hand, but also steadfastness in essence, a high level of social competence, and a great deal of courage." (1) Ever faster-changing business environments require the leadership skills of recent years to change significantly in this often-called "VUCA" environment (Volatile, Uncertain, Complex, Ambiguous). As Sikora prophetically states, "It is naive to believe that the management challenges of the digital age can be overcome with the traditional management methods of the 20th century." (2)

In their article "Digital Reinvention: Unlocking the How," McKinsey & Company consultants state that culture and the associated behavioral changes are the main obstacles to digital effectiveness. (3) They further emphasize that "executives who wait for organizational cultures to change organically will move too slowly as digital penetration grows, blurs the boundaries between sectors, and boosts competitive intensity." In the context of digital transformation, changing culture, attitudes, and behavior are paramount. In fact, a strong top-down direction from the senior executive team, coupled with methods that *actually engage employees in making the change happen,* is the only effective way to drive digital transformation. (4)

Organizations have started to take note—both Digitally Developing, and Digitally Mature companies consider the need for leadership skills to be very high. When Kane was asked whether organizations are effectively developing those "digital age" leadership capabilities, the difference in responses was considerable. While around two-thirds of respondents from *digitally maturing* companies said that they are doing so, only 33% of developing-stage companies and 13% of early-stage companies said the same. (5)

> A strong top-down direction from the senior executive team, coupled with methods that *actually engage employees in making the change happen,* is the only effective way to drive digital transformation.

Defining "Digital Age" Leaders

In this chapter, we take a closer look at digital leadership through the lens of *character* and *competency* in Covey's *Trust Model.* We operationalize his widely accepted leadership framework, strengthen its relevance, and conceptually connect the model to the results of digital transformation. To test the model, we use the results of our international survey *Patterns of Digitization,* which measures the relevant factors organizations currently undertake to transform their businesses. This study informs senior management that *digital transformation is not just about deploying technology; the human aspect of digital leadership—traits, competence, intent, and integrity of a leader—plays a key role in its success* (6).

Patterns of Digitization Survey

With the help of our research partner Institute of Industrial Management FIR at RWTH Aachen University in Germany, we designed, tested (repeatedly with industry partners), and administered an extensive survey instrument, called *Patterns of Digitization,* to assess how global corporations are actually implementing digital transformation. Those of you who have a good memory are probably questioning the title of the survey, *Patterns of Digitization.* Yes, as we covered in Chapter 1, "Trust Is So Important," when we unpacked digital transformation, digitization is just one step—and frankly, not the most important one—in the process of transforming a company to a digital business. At the time we conceived the survey in 2017, we got caught in the same semantics trap of heaping *digitization, digitalization,* and *digital transformation* into one pile. Believe us, the survey focuses on digital transformation and the multitude of factors that must be considered by companies engaged in digital transformation. The survey assesses not only the state of digital transformation in these organizations, including the various strategies they employ to achieve large-scale institutionalized digital transformation, but it also examines the behaviors of its leaders. Specifically, we are looking at how their actions align with the four core values of digital leadership we laid out—*intent, integrity, capabilities,* and *results.*

We broadly distributed this survey and gathered data from 559 decision-makers (middle and senior managers) across five geographic regions—Americas, Europe, Asia, Africa, and Oceania. In the Company Characteristics section, we asked respondents several standard demographic questions, including the region and industry sector their company operates in, and their company's size, age, and number of employees. Nearly 61% of the respondents are from large companies (with over 200 employees), including about 35%

from companies with over 1000 employees. About 37% of the respondents were from companies with fewer than 200 employees including 20% from companies with fewer than ten employees. 60% of the companies that took the survey operated predominantly in Americas followed by 35% in Europe, and the rest from Asia, Africa, and Oceania. Please refer to Figure 5, "Sample Characteristics," to better understand the demographics of the industries.

INDUSTRY SECTOR	Firm Size (No. of Employees)						
	1-9	10-199	200-1000	>1000	UNSPECIFIED	TOTAL	%
IT and Communications	9	15	28	58		110	19.7%
Consulting and Services	33	20	15	16		84	15.0%
Other	25	12	13	12		62	11.1%
Education	6	6	12	16		40	7.2%
Construction and Real Estate	5	11	13	7		36	6.4%
Finance and Insurance	1	7	14	13		35	6.3%
Unspecified	8	3	4	3	13	31	5.5%
Retail and Wholesales	9	5	11	5		30	5.4%
Health Care	4	6	6	8		24	4.3%
Manufacturing	2	2	9	11		24	4.3%
Automotive and Suppliers	3	2	3	9		17	3.0%
Mechanical and Engineering		3	5	8		16	2.9%
Agriculture, Agribusiness and Bioscience	4	3	1	4		12	2.1%
Transport and Logistics	1		4	6		11	2.0%
Chemicals and Pharma		1		7		8	1.4%
Energy and Gas	2		2	4		8	1.4%
Government			3	5		8	1.4%
Industrial Waste			1	1		2	0.4%
Public Administration			1			1	0.2%
TOTAL	112	96	145	193	13	559	
	20.0%	17.2%	25.9%	34.5%	2.3%	100.0%	

Figure 5. *Sample Characteristics*

We also asked participants to rate the digital maturity of their organizations on a five-point scale, from Level 1, Ad hoc ("no formal plan or approach") to Level 5, Optimized ("new business model is fully internalized; results are repeatable and predictable"). Results are shown in Figure 6.

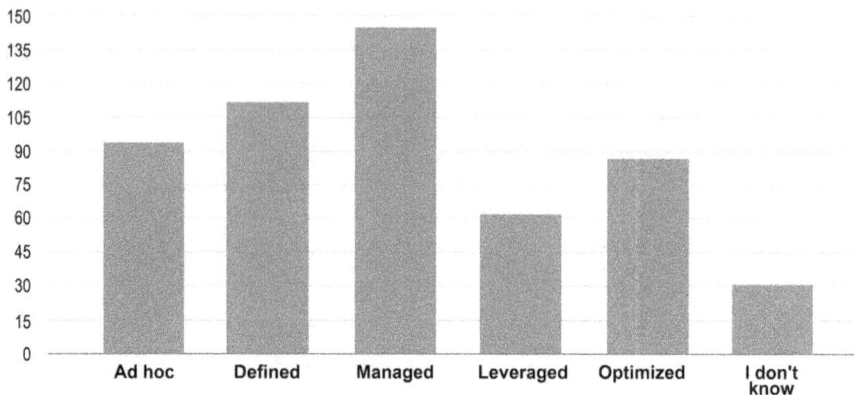

Figure 6. *Digital Maturity Ratings*

Figure 7 shows the top ten predominant industry sectors represented in our survey, along with a breakdown of Digitally Mature and Digitally Developing responses within each sector.

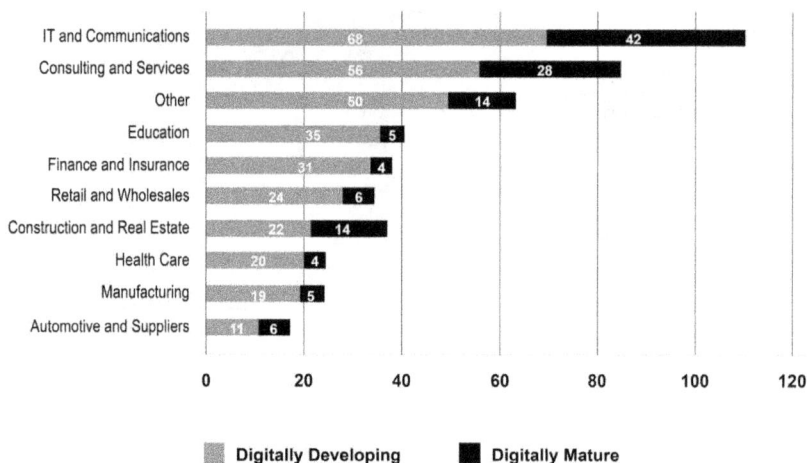

Figure 7. *Top Ten Industry Sectors and Their Digital Maturity*

But how did we calculate which companies and industry sectors were Digitally Developing versus Digitally Mature? More importantly, what does this mean about their leadership?

Digitally Mature vs. Digitally Developing Organizations

We used the Carnegie Maturity Model Integration (CMMI) process to distinguish between Digitally Mature and Digitally Developing organizations. *CIO Magazine* describes CMMI as a "process and behavioral model that helps organizations streamline process improvement and encourage productive, efficient behaviors that decrease risks in software, product and service development." (7) The CMMI model breaks organizational maturity into five levels. For businesses that embrace CMMI, the goal is to raise the organization to Level 5, the "optimizing" maturity level. CMMI considers organizations that achieve Levels 4 and 5 as high maturity— these organizations are "continuously evolving, adapting, and growing to meet the needs of stakeholders and customers." (7) We use CMMI to classify companies into Digitally Mature and Digitally Developing categories. See Figure 8, CMMI Progression.

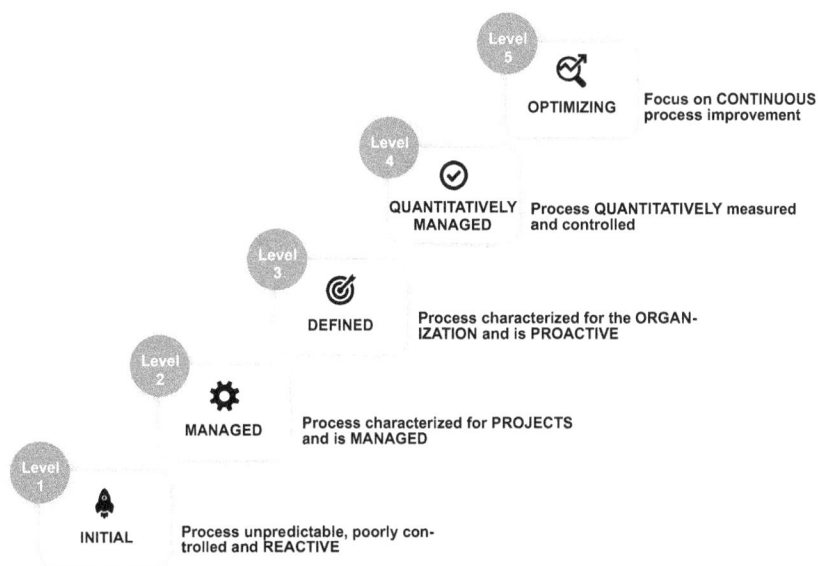

Figure 8. *CMMI Progression*

We classified companies at Levels 4 and 5 as highly mature organizations and labeled them Digitally Mature organizations. Companies operating at Levels 1, 2, and 3 we designated less mature and labeled them Digitally Developing organizations. Using the CMMI process, the literature on digital transformation, and the direct feedback from practitioners, we developed the following definitions for each level of digital transformation maturity in an organization:

❖ Level 5 is the optimized state. Companies operating at Level 5 have achieved digital maturity and are primarily concerned with the continuous improvement of the new business model and business processes.

❖ Level 4 is the level where synergies occur; the company involves competencies and people from outside the organization.

❖ Level 3 is where managers' actions reflect the new, desired behaviors; their goal is to institutionalize the new model.

❖ Level 2 is where organizations make digital transformation a strategic imperative, and a transformation strategy is developed.

❖ Level 1 is the initial state, with no concerted efforts on transforming the organization.

The separation of Digitally Mature and Digitally Developing occurs between Level 4 and Level 3. Level 4 companies have had to involve competencies and people from outside the organization to achieve external initiation strategies, for example, developing an ecosystem, entering a merger and acquisition, or creating a digital spinoff.

By contrast, Level 3 organizations typically concern themselves with improving project performance *with the goal* of institutionalizing the new model, i.e., to eventually internalize it and deal with the inevitable people and cultural issues that surround this next big step. In our estimate, institutionalization may be the hardest task of all, and we believe this is where most Digitally Developing organizations find themselves today. Using these definitions, we determined there were 352 Digitally Developing and 145 Digitally Mature organizations in the survey. (Please refer to Figure 6, Digital Maturity Ratings, for these counts.)

Our premise is that the leaders of the 145 Digitally Mature companies in our survey represent the "best practices" of digital leadership, but what actions best illustrate their practice?

Mapping Covey's Model to the Survey Instrument

To understand these actions, we chose questions from the survey that either directly mention "leaders" (Please refer to Figure 9, Leadership Traits) or could be *inferred* to be the actions of digital leaders. (Please refer to Figure 10, Design Philosophy and Communications, and Figure 11, Strategic Investments.)

Figure 9 lists six digital leadership questions from the survey. We aligned these questions with the major categories of *character* and *competence* in Covey's model. Please see Figure 9, Leadership Traits.

	ITEM	CATEGORY
1	Our leaders act and behave (walk-the-talk) as promoters for the digital transformation process	Character
2	Our leaders promote 'fail fast' culture which helps employees to learn from mistakes	Character
3	Our leaders think like entrepreneurs and promote this mind-set actively to their employees.	Competence
4	Our leaders make decisions less on intuition or experience but more by utilizing facts and analytics	Competence
5	Our leaders represent extensive digital technology expertise	Competence
6	Our leaders collaborate with cross-functional business counterparts	Competence

Figure 9. *Leadership Traits*

Figure 10, Design Philosophy and Communications identifies five questions that were determined to be predominant practices of Digitally Mature organizations. These practices address the so-called softer factors of digital transformation and organizational culture. And while these questions do not mention the word leader, we think these are the result of the *capabilities* of modern, competent digital leaders.

	ITEM	CATEGORY
RAPID PROTOTYPING		
1	We develop our products and services by developing prototypes that we test with customers and partners and learn from it (e.g. Rapid Prototyping)	Capabilities
DESIGN THINKING		
2	We develop utilizing interactive teams' ad-hoc ideas for new products and services by taking a customer's perspective (e.g. Design Thinking)	Capabilities
OPEN COMMUNICATION		
3	Our project management approach utilizes agile principles, by communicating progress and barriers transparent and regular to all project members (e.g. Open communication)	Capabilities
AGILE DEVELOPMENT		
4	Project results are iterative and in short sequences. We react quickly to changing requirements (e.g. Development sprints, Scrum)	Capabilities
OPEN INNOVATION		
5	We generate new ideas for products and services by open communication and idea exchange with customers, suppliers and business partners (e.g. Open innovation)	Capabilities

Figure 10. *Design Philosophy and Communications*

Figure 11, Strategic Investments, shows the seven survey questions associated with the investments made by the organizations. We have categorized all of them under Covey's core value of *results*. Nothing could be more important than these visible actions to raising employees' trust in the organization's transformation strategy—and its digital leader.

	ITEM	CATEGORY
1	Forming a Taskforce	Results
2	Hiring Consultants	Results
3	Appointing a CDO (Chief Digital Officer)	Results
4	Developing a Data Strategy	Results
5	Hiring or Training a significant number of Data Scientists	Results
6	Moving one or more of our products/services to the cloud	Results
7	Establishing one or more new touch points with customers	Results

Figure 11. *Strategic Investments*

Respondents were asked how often these 18 actions (in Figures 9-11) are observed: 0% = Never, 25% of the time, 50% of the time, 75% of the time, 100% = Always, and I don't know).

Now let's look at the data.

What We Learned About Digital Leaders

Character

Our findings confirm and underscore the importance of the *character* and *competencies* of digital leaders. Leaders of Digitally Mature companies have high *integrity*; they definitely "walk the talk" (much more than the leaders of Digitally Developing companies). These leaders are *promoters* of digital transformation (see Figure 12 below).

It is interesting to note that a leader's integrity is biggest difference between the two groups. According to Covey, integrity is the root of leadership, and our data emphasizes its importance.

Our leaders walk-the-talk as promoters for the digital transformation process

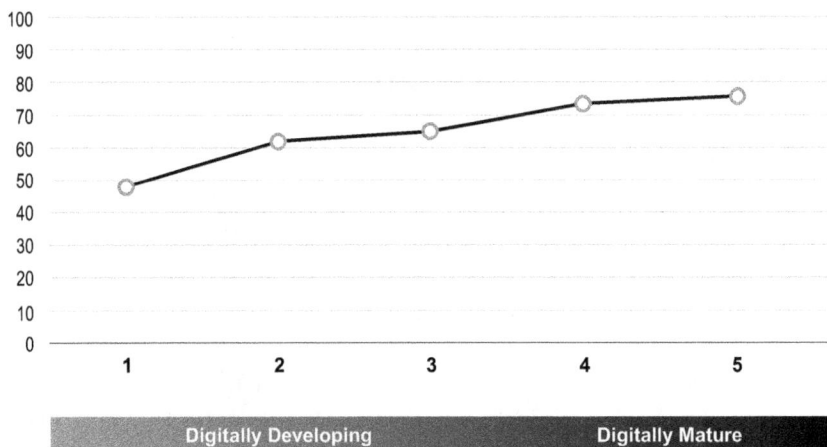

Figure 12. *Promoters of Digital Transformation*

Explanation of the graphics

Before we go any further, let's understand what the graphics are telling us.

The x axis shows the spread of Level 1 to Level 5 organizations. Remember, Digitally Mature organizations are Level 4 and Level 5.

The y axis is the percentage of the time these actions are observed. In this case, the range of responses was from 49% to 75%; the resulting delta is $\Delta_{1,5} = 26\%$. Said differently, the response "Our leaders act and behave (walk-the-talk) as promoters of the digital transformation process" is observed 49% of the time in organizations with a digital maturity rating of 1, versus 75% in organizations with a digital maturity rating of 5.

Along the x axis, you can see the percentage of the time this response was selected by Level 2, 3 and 4 organizations. We did not include these data points on subsequent graphs. We just want to show you how the slope of the line is generated. In this case, the slope of the line was almost a straight line up and to the right, showing that this behavior progresses positively across the maturity levels.

We also calculated the statistical difference between Digitally Developing (352 companies) and Digitally Mature (145 companies) companies, which is significant in this case with a "medium to strong effect strength" ($t_{(415)}$ = 5.784, $p < 0.01$, $d = 0.639$).

We are not going to go through the statistical calculations that support these differences each time. We will stick to just reporting the range and slope of the results. If you would like to understand the rigor we applied in mapping these actions, please read the article we authored, "DIGITAL LEADERSHIP: Character and Competency Differentiates Digitally Mature Organizations" (8). The article, written for the 2020 IEEE International Conference on Engineering, Technology and Innovation (ICE/ITMC), is the first time we reported that the digital position of a company—whether it is digitally mature or still developing its position—depends on the trust-building actions of its leaders.

Our leaders promote fail fast culture which helps employees learn from mistakes

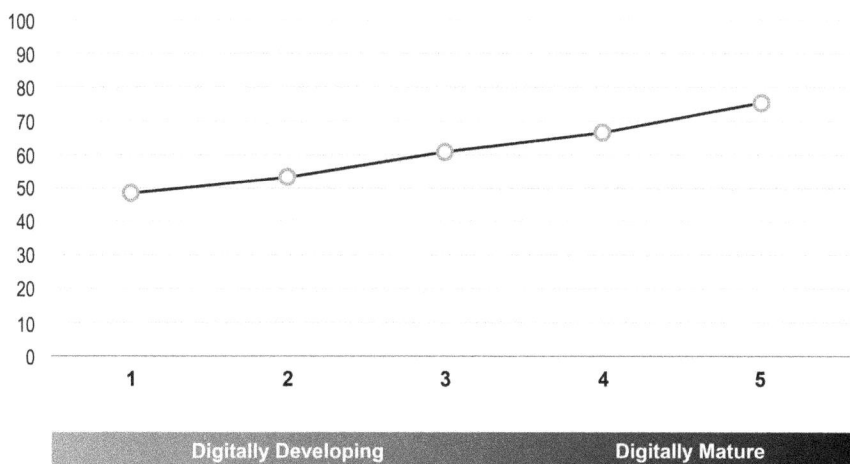

Figure 13. *"Fail Fast" Culture*

Leaders of Digitally Mature organizations also promote a "fail fast" culture which ups the pace of digital transformation as well as helping the teams learn from their mistakes. Please see Figure 13. "Fail Fast" Culture. This way everyone is on the same page as to what's important, i.e., the *intent* of the transformation strategy and what they need to do better and faster. The range of these responses is 48% to 76%.

Competencies

Leaders of Digitally Mature organizations are *competent*. Please see Figures 14, 15, 16, and 17 below and the substantial range of results in all cases.

These leaders are knowledgeable, exhibit entrepreneurial behaviors, and promote digital transformation. The behaviors of these leaders directly impact employee performance and therefore affect the implementation of digital transformation strategies. According to our results, digitally mature companies were much more likely to have leaders with extensive technological expertise and to make decisions less from intuition and more based on data and facts.

Our leaders represent extensive digital technology expertise

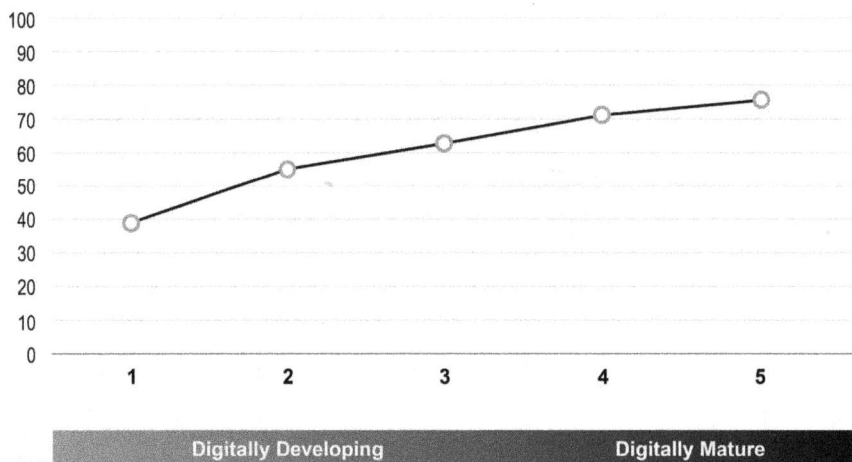

Figure 14. *Digital Technology Expertise*

54

Our leaders utilize facts and analytics to make decisions

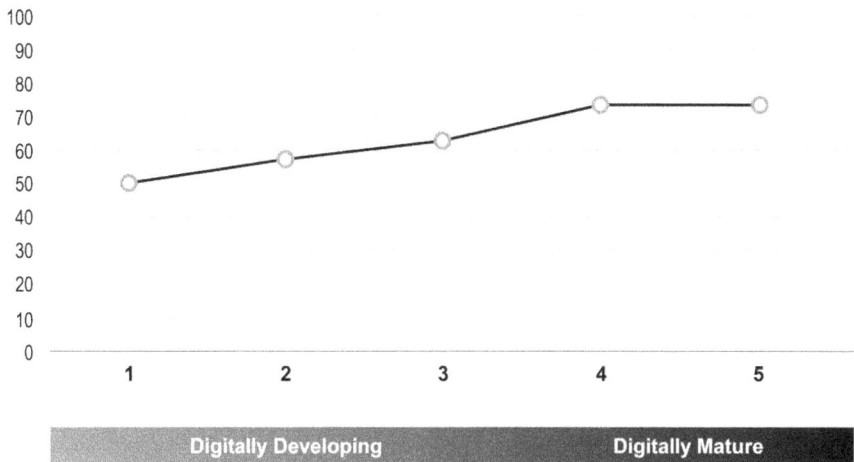

Figure 15. *Utilize Analytics*

Our leaders promote entrepreneurial mindset

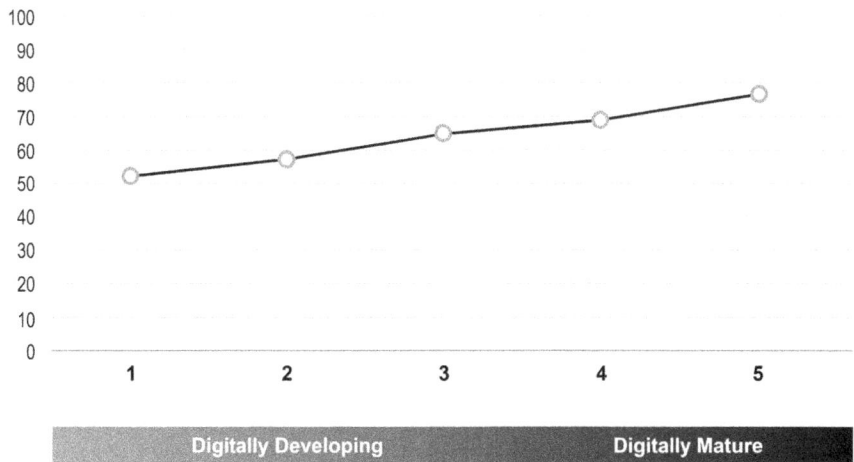

Figure 16. *Entrepreneurial Mindset*

Our leaders collaborate with cross-functional business counterparts

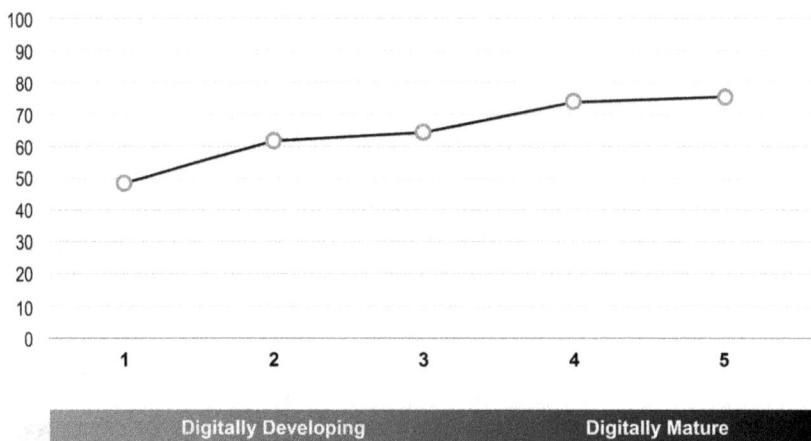

Figure 17. *Collaboration*

Right here, based on their character and competencies, we have an excellent "prescription" for a digital leader. But let's go deeper into the competence category of Covey's model and assess these leaders' *capabilities* and *results*.

Capabilities

Leaders of digitally mature organizations are *capable*. Please see Figures 18, 19, 20, 21, and 22 below. Our data shows that these digital leaders leverage AGILE project management philosophy and work collaboratively with their peers. We also found that digital leaders foster timely and open communications. There is a sizeable disparity between Digitally Mature and Digitally Developing organizations in their level of open communication. Specifically, our results show that digitally mature organizations have higher amounts of transparent and open communication across their organizations. Of course, this behavior is expected, if not outright demanded, by their leaders.

We develop products and services by developing prototypes that we test with customers and partners and learn from it (e.g. Rapid Prototyping)

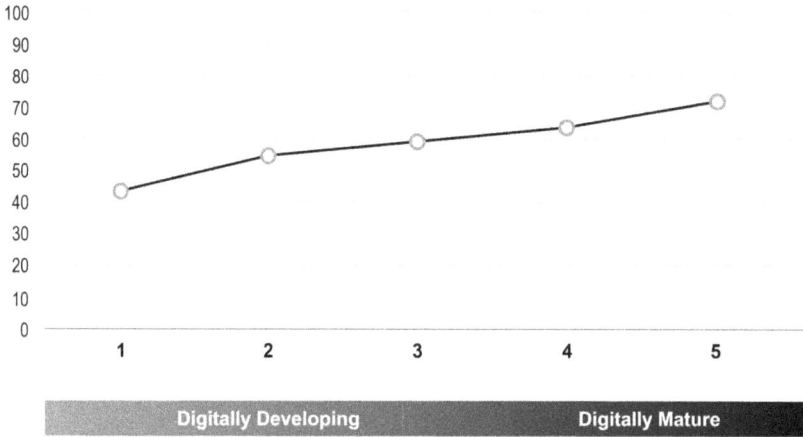

Figure 18. *Rapid Prototyping*

We utilize interactive teams' ad-hoc ideas to develop new products and services by taking a customer's perspective (e.g. Design Thinking)

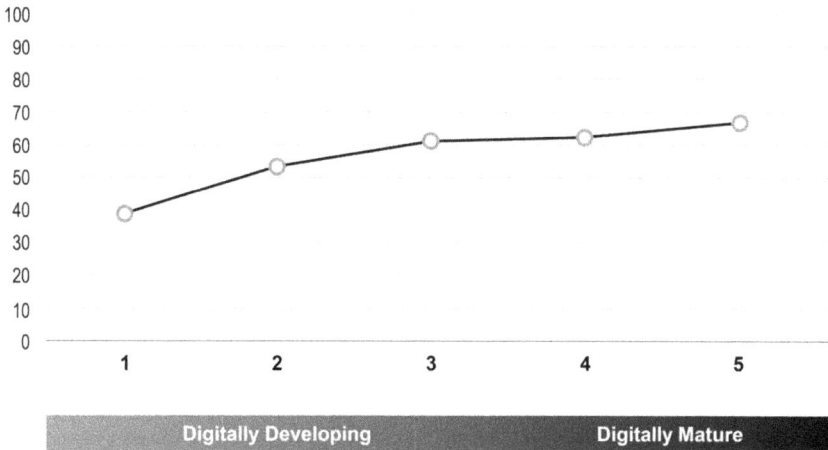

Figure 19. *Design Thinking*

Our project management approach utilizes AGILE principles and open communication

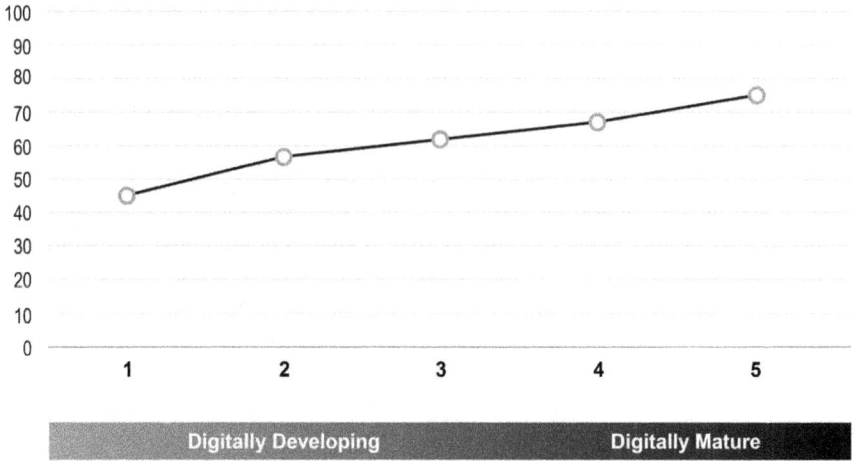

Figure 20. *Open Communications*

Project results are iterative and in short sequences. We react quickly to changing requirements (e.g. Development sprints, Scrum)

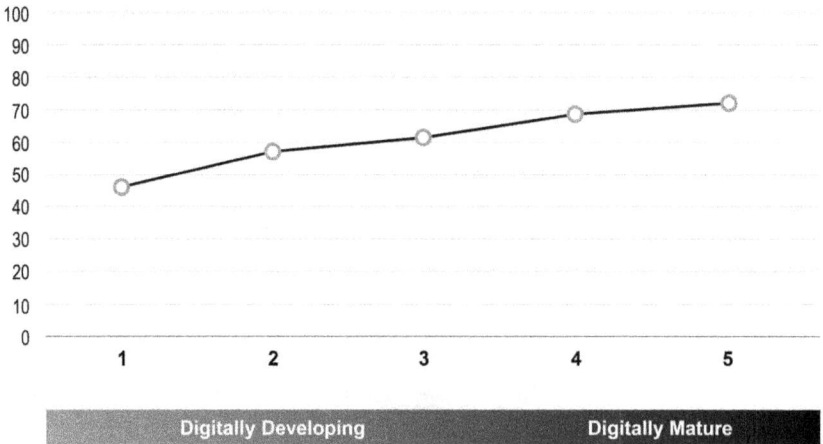

Figure 21. *AGILE Development*

We generate new ideas for products and services by open communication and idea exchange with customers, suppliers, and business partners (e.g. Open innovation)

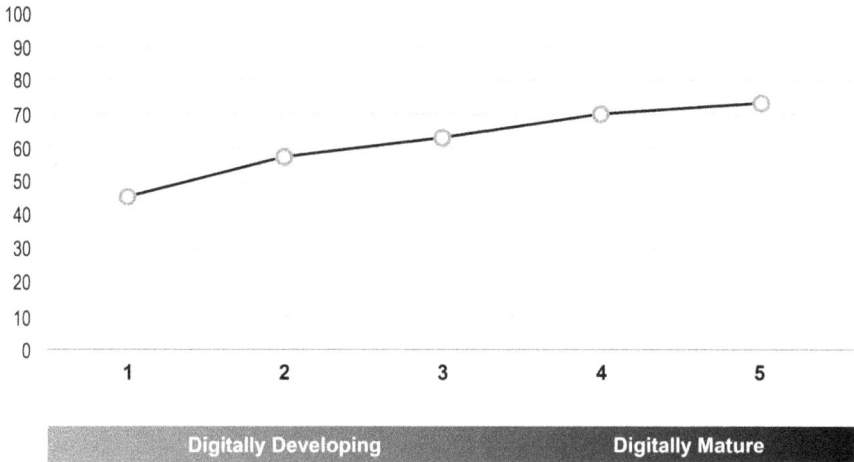

Figure 22. *Open Innovation*

The leaders of digitally mature companies employ rapid prototyping, design thinking, and AGILE development methods, and they engage in open innovation. I think you'll agree these capabilities are quite impressive and are the result of strong, capable digital leaders.

Do we think that observing leaders exhibiting these modern capabilities would build confidence in employees, suppliers, and business partners about the organization's ability to transform itself?

Of course, we do—but we are not yet done.

Results

Establishing a visible track record is essential to the success of digital leaders. From the data, we can see that the leaders of digitally mature organizations *always* align resources with their strategy. To test this premise, let's examine the strategic investments these leaders are making. Please see Figures 23, 24, 25, 26, and 27 below.

If you look at Figure 11, Strategic Investments, you will see that the survey assessed seven different actions. We are not going to talk about action number 1, "Forming a Task Force" or action number 2, "Hiring Consultants." These are the actions of companies—and their leaders—who are just getting started. The leaders of digitally mature organizations should be far beyond this point.

We could say that developing a data strategy is "jacks or better in this poker game." You can see from the data that survey respondents observed this behavior about evenly from Level 2 through Level 5. If you recall the CMMI definitions of these levels, this makes sense. Only Level 1 companies, which have "no concerted effort in digital transformation," fail to score high on this action. Everyone else, Levels 2 through 5, needs to "put a premium on data" and develop their data strategy. (Please see Figure 23.)

Appointing a chief data officer (CDO) was one of actions least taken overall, but our respondents indicated it is vital for a digitally mature organization. (Please see Figure 24.) Note that the slope of performance is not linear and "up and to the right." Only Level 5 organizations stood above the rest on this action.

Hiring and training data scientists is another important action. From the data you can see that only the leaders of Level 4 organizations excelled at this action. (Please see Figure 25.) Nevertheless, the leaders of digitally mature organizations outperformed Digitally Developing organizations.

Last and certainly not least, the leaders of digitally mature companies "moved their products and services to the cloud." (Please see Figure 26.) In this case, the digital leaders of Level 5 companies were the top performers. This logically fits with organizations operating at the optimized level of maturity, doesn't it? Again, the leaders of digitally mature organizations at Levels 4 and 5 outperform those of still developing organizations.

We consider this last action, "established new touch points with customers *electronically,*" as the acid test of the digital leader's ability to build trust. This action signals the company's business model has been rewired. All the other actions, while important, were preparatory. The change is real now for all employees. As you can see from the data, the top performers were digital leaders of Level 4 organizations. (Please see Figure 27.)

Developed a Data Strategy

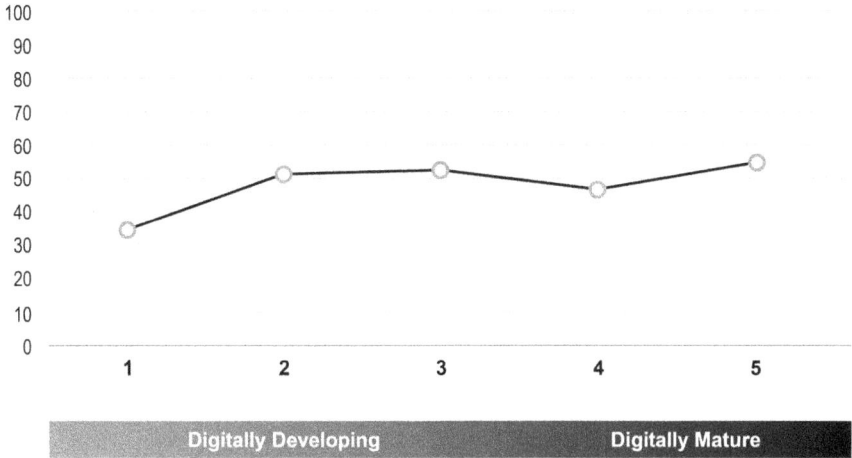

Figure 23. *Developed a Data Strategy*

Appointed a Chief Digital Officer

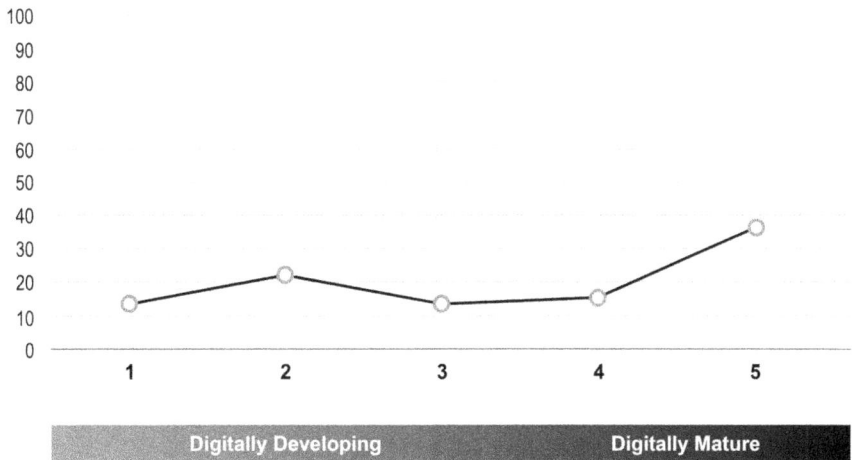

Figure 24. *Appointed a CDO*

Hired or trained a significant number of Data Scientists

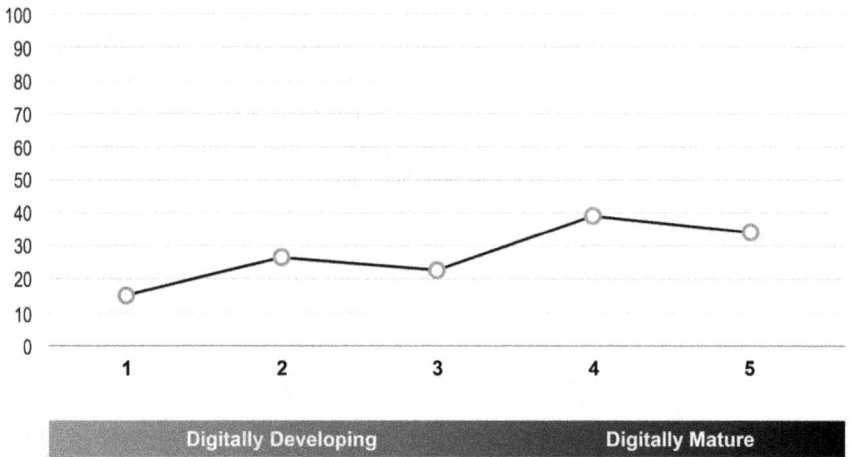

Figure 25. *Integrated Data Scientists*

Moved one or more products/services to the cloud

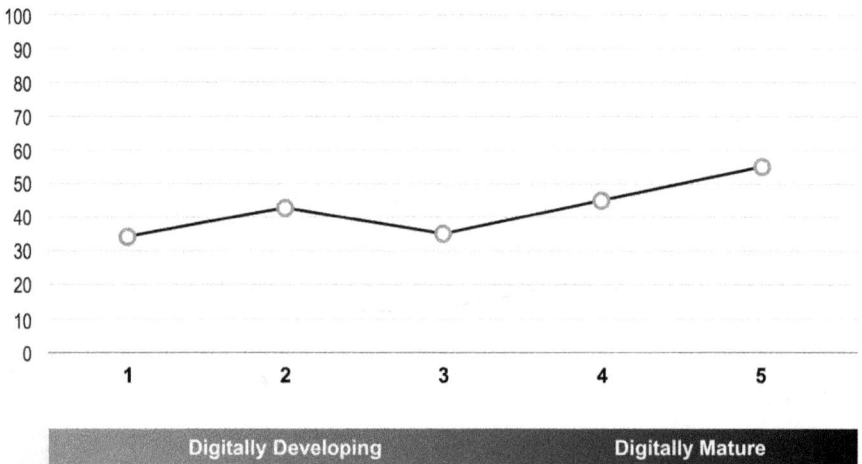

Figure 26. *Moved Product/Service to Cloud*

Established one or more new touch points with customers electronically

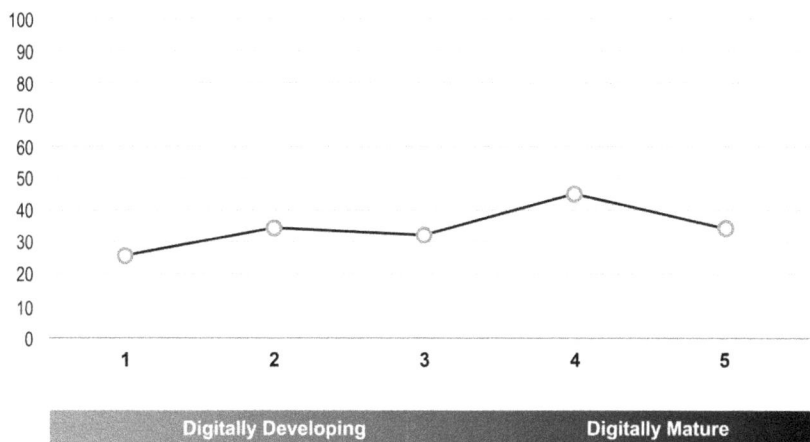

Figure 27. *Established Electronic Touch Points with Customers*

By these measures, we can see that the leaders of digitally mature organization produce *results*. Covey highlights that leaders need to demonstrate results—nothing creates trust as quickly as solid results.

THE Result: Units Adopt Transformation

The Patterns of Digitization survey also assessed whether business units had adopted the transformation strategy. This is really the ultimate test of digital transformation, isn't it? Figure 28, Characteristics of Functional Units, identifies the key actions of line units of the companies identified in the survey. These actions are those of employees "looking up" at the efforts of their leaders to transform the business. In the case of these questions, respondents answered "yes" to them over 65% of the time.

	ITEM	CATEGORY
1	Was your functional group established for the express purpose of enabling a digital transformation?	Results
2	Are the organization's KPI (Key Performance Indicators) tied to the attainment of these capabilities and resulting benefits to the organization at large?	Results
3	We actively invest in training employees to develop the knowledge, skills, and abilities necessary to survive in the Digital Transformation	Results
4	Are major market trends (Political, Economic, Societal, Technological, Environmental and Legal) routinely monitored and shared with employees?	Results
5	Does the strategy include acquiring or developing new digital capabilities?	Results
6	Is the strategy broadly communicated to department employees?	Results
7	Does your function have a strategy which addresses the opportunities and threats inherent in these trends?	Results

Figure 28. *Characteristics of Functional Units*

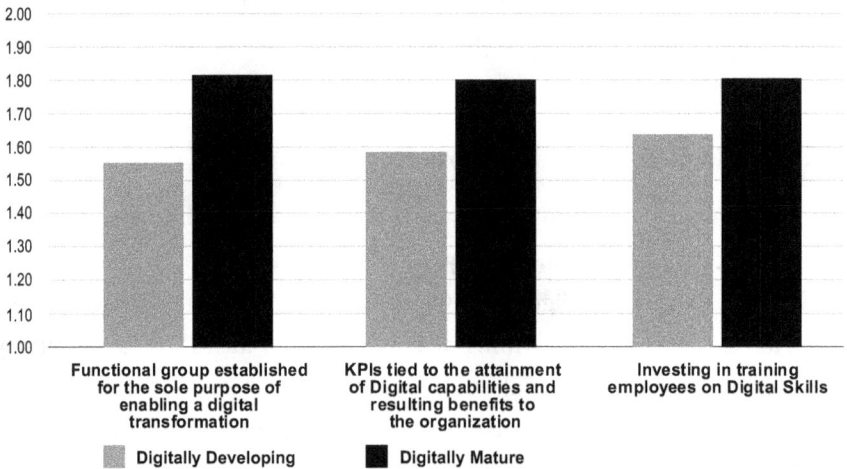

Figure 29. *Functional Units Adopt Transformation*

When we then plotted these practices against their digital maturity levels, to no one's surprise, digitally mature organizations dominated these practices (see Figure 29). In this case, a "no" answer was given a 1; the answer "yes" was given a 2. As you can see, Digitally Mature organizations scored 1.8 or higher for each question. These results should be reaffirming to the digital leader. The strategy they formulated, the investments they made, even how they communicated to the organization is paying off. The units have bought in and are amplifying the plan.

Final Thoughts

In this chapter, we established an important connection between Covey's Trust Model and the actions of digital leaders. Using a dichotomous categorization of Digitally Developing and Digitally Mature companies, we structured and scored the results from the *Patterns of Digitization* survey. Results confirm that Digitally Mature companies are more ambitious about digital leadership than Digitally Developing companies. Leadership seems to be approached proactively by Digitally Mature companies and is understood as an important key component for future success. Both their character and competency motivate digital leaders to build trust and credibility and to take differentiated actions to set their companies apart.

> Leadership seems to be approached proactively by Digitally Mature companies and is understood as an important key component for future success.

McKinsey argues that trust must prevail at all levels of the company, regardless of industry, to be able to take new risks in the digital environment. (3) Without this trusting leadership mindset, all aspects of digital leadership will succeed only partially or become a meaningless duty exercise for employees. (10)

The specific lists of differentiated actions we identified should give you a work plan—a winning formula, if you will—as you embark on your own digital transformation journey. Moreover, if you want to be a digital leader and understand what we have learned directly from digital leaders about how they engender trust, please read the rest of the book.

Chapter References

1. Teichmann, S. and Hüning, C. (2018). *Digital Leadership – Führung neu gedacht: Was bleibt, was geht?* in *Disruption und Transformation Management: Digital Leadership – Digitales Mindset – Digitale Strategie,* F. Keuper, M. Schomann, L. I. Sikora, and R. Wassef, Eds., Wiesbaden: Springer Fachmedien Wiesbaden. pp. 23–42.

2. Sikora, H. (2017). Digital Age Management: Führung im digitalen Zeitalter. *Elektrotech. Inftech.*, 134(7), 344–348.

3. Bender, M. & Wilmott, P, eds. (2018). *Digital reinvention: Unlocking the 'how'.* Digital McKinsey.

4. Westerman, G., Bonnet, D., & McAfee, A. (2014). *Leading digital: Turning technology into business transformation.* Boston, Massachuchetts: Harvard Business Review Press.

5. Kane, G.C., Phillips, A.N., Copulsky, J., & Andrus, G. (2019). How Digital Leadership Is(n't) Different. *MIT Sloan Management Review.*

6. Mugge, P., Abbu, H., Michaelis, T.L., Kwiatkowski, A., & Gudergan, G. (2020). Patterns of Digitization: A Practical Guide for Organizations Engaged in Digital Transformation. *Research - Technology Management,* 63(2), 27–35.

7. White, S.K. (2018). What is CMMI? A model for optimizing development processes. *CIO,* March.

8. Abbu, H., Kwiatkowski, A., Mugge, P., & Gudergan, G. (2020). DIGITAL LEADERSHIP - Character and Competency Differentiates Digitally Mature Organizations. IEEE International Conference on Engineering, Technology and Innovation (ICE/ITMC).

9. Ready, D., Cohen, C., Kiron, D., & Pring, B. (2020). The New Leadership Playbook for the Digital Age. *MIT Sloan Management Review.*

CHAPTER 4

LISTENING TO SUCCESSFUL LEADERS

"The distance between number one and number two is always a constant. If you want to improve the organization, you have to improve yourself and the organization gets pulled up with you."

– Indra Nooyi

Recap

Previously we unpacked the major steps involved in digital transformation to understand how this phenomenon evolves. We recognized the very real fears companies see as they worry about committing their own organizations to what may seem like very perilous journeys. Most importantly, we put our stake in the ground that *the difference between companies that succeed at digital transformation and those that don't is embodied in a very human trait of their leaders—how they embrace and enable trust* (Chapter 1, "Trust is So Important").

In Chapter 2, "Science of Trust," we summarized what learned researchers tell us about trust and the important function it plays in leadership. As part of this overview, we did a deep dive into Covey's Trust Model. We were particularly taken with his model because its whole premise is that "high trust" organizations can accomplish change more quickly—and at less cost—than low trust organizations.

In Chapter 3, "Digital Maturity Demands Digital Leadership," we presented the result of a global research study we did into the practice of Digitally Mature and Digitally Developing organizations. We were particularly interested in how leaders of the Digitally Mature firms measure up on Covey's four core values of credibility: *integrity, intent, capabilities,* and *results*. The net is they do…strongly.

Now we want to talk to digital leaders directly to understand how, exactly, they build trust across their organizations to lead a digital transformation.

How We Designed the Interview Guide

The basis for the interview instrument is Covey's Trust Model. Again, we use elements of Covey's model, which shows all the subdivisions of the four core values as the framework for the interview guide. (Please see Figure 30, Digital Leadership, extended model.)

Digital Leadership

Character — Competency

Integrity — Intent — Capability — Results

Integrity:
- Honesty
- Congruence
- Humility
- Courage

Intent:
- Motive
- Agenda
- Behavior

Capability:
- Talents
- Attitudes
- Skills
- Knowledge
- Style

Results:
- Past
- Present
- Future

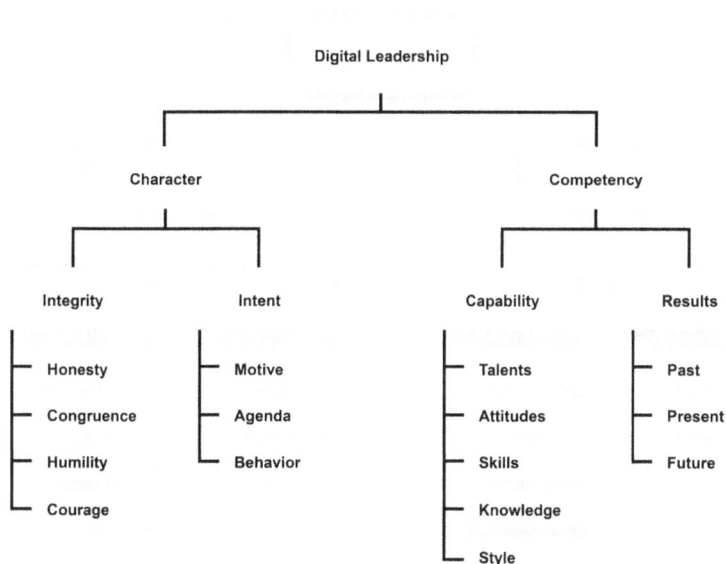

Figure 30. *Digital Leadership Extended Model (Adapted from Covey's Trust Model)*

If you would, please now go to the **Appendix, "Interview Guide."** You will see how we used the model to develop the questions for digital leaders. We have at least one question for each subdivision of the four cores.

As you look at the questions, you will see the influence of Covey's 13 Behaviors of High Trust in their design. For example, if we look at the first question…

> 1. *(Honesty) How often do you publicly address employees' concerns with the strategy? Do you insist employees be transparent and talk straight as well?*

You can see how this not only aligns with the first behavior, *talk straight* (please refer to Chapter 2, "Science of Trust" again), but it also goes further. We assume these successful leaders *do* talk straight; we are asking them if they insist this behavior be repeated across the organization. This is how they can better build relationships and develop organizational trust.

While you are in the Appendix, why don't you attempt to answer the questions yourself? This will give you a much better feel for what we are looking for, and you may learn some interesting facts about your own ability to generate trust.

How We Picked the Leaders

We developed a database of the leaders we wanted to interview. The selection factors that we considered were

- *Geographic representation:* We wanted a good mix of regional representation, primarily from the US and Europe.
- *Size of the company:* We are not just interested in what the leaders of established firms are doing but also what their counterparts at small and midcap businesses are experiencing. Is it harder or easier for them to build trust?
- *Gender:* We want to know what it is like for female digital leaders to build trust and change culture at primarily male-dominated organizations. We are sure it is harder, but what are they experiencing? We want to know.
- *Cross-industry representation:* We also want to take a broader view of digital transformation than, frankly, a lot of business articles portray today. The tend to focus on "digital-native" firms, which is natural. We want to look at those as well, plus more traditional "smokestack" industries, including automobile, chemical, farming, healthcare, and heavy machinery manufacturing. These companies don't receive the press that digital-native firms like Google and Microsoft do—but they are just as important.

Knowing we would have only a short time with these busy people, we chose 20 questions to cover the four core values of *intent, integrity, capability,* and *results.* Within each of these core values, we can probe deeper. (Please see

Interview Guide.) For example, within *integrity* we have questions for *honesty, humility, courage,* and *congruence.* But even this is too much for a single 60-minute interview.

To solve this, we gave the interviewee the opportunity to choose questions from two themes: a primary theme like *integrity* under *character* and a secondary theme from the *competence* category, like *capabilities* or *results.* Thirteen interviews should give us good coverage of our digital leadership thesis while limiting and condensing the time with any one person.

Layout of Interview Reports

We did our homework. Before we conducted an interview, we developed a biographical profile of the leader, including a summary of their organization. That way, we could get right into the questions when we met with them. These personal interviews—and the lessons learned from them—are the heart of the book!

We are not trying to have each interviewee answer each question. Rather, we want to generate a rich dialogue into each of the topics. We summarize these in Chapter Six, "Winning Formula," so readers can get a deeper understanding of their leadership behaviors.

Again, the interview is not a test; we *know* these people to be successful digital leaders. What we are trying to ascertain is *how* they and their organizations have achieved their current position. Specifically, we want to know how their trust-inducing actions influenced the speed and cost of transformation.

CHAPTER 5

THE PULSE OF DIGITAL LEADERS

"The day soldiers stop bringing you their problems is the day you have stopped leading them. They have either lost confidence that you can help them or concluded that you do not care. Either case is a failure of leadership."

– General Colin Powell

Digital Leaders We Selected

The essential part of this book is the pulse of digital leaders. We were fortunate to talk with a motivated and diverse group of digital leaders and hear them—straight and unfiltered—discussing their successes, regrets, and the lessons they learned in doing the hard work of digitally transforming their companies. They share their and their organizations' experiences and their messages to other digital leaders.

Again, these people were not randomly picked; they were purposefully selected. These global digital leaders—of both genders—represent organizations of all sizes, one-person micro companies to mega cap giants with a half million employees, small family businesses to large public companies. More importantly, they represent "more traditional" industries—automobile, agriculture, manufacturing, business services, and others—the ones that are aspiring to catch up with their digitally native brethren. Frankly, publishing their stories is more important, because there are *a lot more* traditional companies than there are digital natives. We think the leadership stories from these people are every bit as interesting, and probably more useful, to the readers of this book.

In Chapter 6, "The Winning Formula," we go deeper into these people's behaviors and actions and summarize the lessons they've learned, and hopefully, you will follow them to *make trust the centerpiece of your strategies*. You can certainly jump ahead and read the most interesting pieces, but please take the time to read the entire set. Before you do, let us share with you the process we used to capture their ideas.

Interview Process – Capturing the Voice of the Leaders

To the best of our ability, we tried to capture the thoughts and actions of the leaders *in their own voices*. In the first four chapters of the book, we tried to put into prose the case for why digital leaders should build trust to successfully lead a digital transformation. Chapter 5 is different. Here we acted more like editors, trying to extract and weave into cogent stories the beliefs and actions of these leaders. We think capturing the main messages in their own words is particularly important.

In the United States and Germany, we had at least two interviewers on each call so we could double check what we heard the leaders tell us. To make sure we had the right amount of actionable information, we asked the leaders a core set of questions. (Please refer to the Appendix: Interview Guide.)

1. *What is your key theme (main message) for transforming the organization?*
2. *What is your organization's digital strategy and vision?*
3. *What actions have you taken to achieve your theme?*
4. *What results have you witnessed to date?*
5. *What advice do you have for other digital leaders?*

We recorded each interview session and created drafts of what they told us. And, of course, we reviewed the interviews with each leader to make sure we captured their thoughts correctly.

In Chapter 6, we revert to our previous style and collate the numerous inputs we received into what we call the "winning formula for digital leaders." For now, though, put that aside and read who we interviewed and what they told us...

Interview Summaries

- **5.1 Chuck Sykes, CEO, Sykes Enterprises:** *"Look Outside to Re-Assess Your Business Model First"*

Chuck provides some sage advice to executives when transforming their companies. He advises leaders to start on the outside—go back to the "top of the mountain," check and scan the marketplace, and assess whether the business model needs to change. Don't start by focusing on how you can better deliver what you currently do; step back and look at the whole business model and ask, *Well, what do we sell and who do we sell it to?* Also, see what Chuck has to say about the importance of integrity and intent when leading a major change program.

- **5.2 Andera Gadeib, CEO, Dialego:** *"Shape Digital Proactively!"*

This experienced CEO and successful entrepreneur gives us some great insights into the courage and determination it takes to lead a digital transformation. As part of her message, Andera candidly addresses the deficiencies that exist in educating future data scientists. The message is this: A lot of the development of your future workers is going to fall on you, so pay particular attention to the people you employ—like Andera always does.

- **5.3 Larry Blue, CEO, Bell & Howell:** *"Fueling a Startup Culture and Entrepreneurial Mindset"*

Larry addresses what has to be one of the toughest challenges when transforming an established business: culture. He believes in starting small and building from there. Don't try changing everything all at once—success breeds success. People will see what you are paying attention to. They will want to be part of a successful team, and that will allow the culture to change naturally. As he says, it is not about getting it right the first time; it's about how well we learn from our mistakes. If you respect the individual and create an environment where creative, data-driven exploration is nurtured, you'll be able to build an innovative culture. From their results, you will see that Bell & Howell is well on its way to creating a new culture and a successful digital business.

- **5.4 Robert Kallenberg, Director, Strategy and Organization, Porsche AG:** *"Much More Is Possible than Anybody Thought Before"*

Robert describes witnessing the significant digital transformation of the automobile. He asserts that at Porsche, there is no longer a digital strategy and a business strategy—they are one and the same. When Porsche first started its digital journey, it was characterized by a lot of "experimentation." Now Robert is convinced they can achieve more than they ever believed possible. He implores digital leaders to develop a "culture of trust" and to take a hands-on approach to learning new digital methods—these will serve you well.

- **5.5 Brandon Batten, Owner and Operator, Flying Farms LLC:** *"The Trials of Launching a New Digital Business for Family Farmers"*

Brandon is a co-manager of a 24x7, family-owned farm. At the same time, he is creating an all-new technology business for farming. Please read about his exciting applications of drones and digital imagery. Brandon likens family farming to a "handshake business." Do you think trust is important to these people? Please see the actions he takes to continually earn the trust of his clients—and his family.

■ **5.6 Marc Schlichtner, Principal Key Expert and Founding Member of the T-Club, Siemens Healthineers:** *"Transformation 'Evangelists' Are Vital!"*

Through his vast experience as a senior executive, Marc offers us a set of issues that senior management need to deal with, like confiding that the managers may have "blind spots" or adjusting the incentive system, so it is more meaningful to top technical performers. Marc spends much of his time evangelizing digital transformation. We think you will find particularly interesting how he describes the transformation process at Healthineers as both a "bottom-up" and "top-down" process and the actions he takes to balance them.

■ **5.7 Seth Kaufman, President and CEO, Moët Hennessy, North America:** *"Soft Skills from a Leadership Perspective Can Be the Difference-Maker"*

Seth told us that the success of digital comes down to people and the leadership team's emotional intelligence and soft skills. It's about speed, agility, empowerment, humility, empathy, and trust. As he says, when you start from a place of people first, empowerment automatically becomes part of the leadership model for which trust is essential.

■ **5.8 Deborah Leff, Former Global Leaders and Industry CTO of Data Science and AI, IBM:** *"Stop Experimenting with AI—Scale It!"*

Deborah, in her advice to digital leaders, implores us to not only ensure the team using AI plays fair but the machine models that generate recommendations are also "fair" and unbiased. Fairness has to be a conscious endeavor, and it's the responsibility of the C-level officers and the board to know that these models, which will drive the next generation of what we do, are indeed fair.

■ **5.9 Krishna Cheriath, VP, Head of Digital, Data and Analytics, Zoetis Inc. and former CDO, Bristol Myers Squibb:** *"Be Flexible, Learn to Lead from the Front—and from Behind—When Transforming Large, Established Organizations"*

Krishna explained to us how on many days he feels like an "accordion," meaning he needs to be flexible and match the digital services he offers to the abilities of various business functions he serves. The examples he gives are great advice for readers dealing with large, diverse organizations that have very different needs and objectives. Naturally, he cites the "trust and transparency" he establishes with people as the reason for his success.

■ **5.10 Dominik Schlicht, CEO, Talbot New Energy AG:** *"Listening Is the Key for Transformation"*

Dominik puts honesty and having a positive attitude as the most important traits for building trust and succeeding at digital transformation. We think you will like the examples he gives us, working with suppliers, employees, and business partners to achieve impressive results. And, most importantly, he begins all these initiatives with another very powerful—but quiet—action: he listens.

■ **5.11 Craig Melrose, Executive Vice President, Digital Transformation Solutions, PTC:** *"Digital Accelerates and Amplifies Traditional Metrics of Running a Business"*

Craig hits Covey's thesis that "trust decreases cost and accelerates speed" head on. Please note how PTC is generating double digit results for both. Also read how he personally makes himself "vulnerable"—to both his employees and to customers—to raise his personal trustworthiness in their eyes. Can you imagine seeing *vulnerable* on executive competency models in the future? Craig can.

- **5.12 Dagmar Wirtz, CEO, 3WIN:** *"Make Ideas Happen—Openness, Honesty, and Innovation Drive Success"*

Dagmar provides good examples of how to motivate employees to be open and develop their ideas. While some of her message might sound a bit like tough love—for example, she demands transparency and honesty—you will enjoy how she personally selects and develops new ideas. And as you might expect, to be successful, this very accomplished entrepreneur spends an inordinate amount of her time listening to and caring for her workers.

- **5.13 Rahul Basole, Managing Director and Global Lead for Visual Data Science, Accenture AI:** *"AI-Powered Enterprises Require Visual Analytics"*

Rahul provides some interesting examples of this relatively new analytics technology. More importantly, he explains that to gain the trust of organizations, he has to convince decision-makers that he is giving them something that traditional reporting and dashboard tools could not do before. Like other digital leaders, Rahul emphasizes the importance of storytelling. As he points out, it is not just about understanding the problem and perhaps the solution but also about communicating both effectively. Visual analytics can be part of your storytelling toolkit.

CHAPTER 5.1

Look Outside to Re-assess Your Business Model First

CHUCK SYKES
CEO, Sykes Enterprises

Profile

Chuck Sykes was appointed President and Chief Executive Officer of Sykes Enterprises in August 2004. During his 15 years of leadership, he has helped Sykes become a leading global digital marketing and customer care outsourcer. Chuck has nearly three decades of experience at Sykes in all aspects of the business, including operations support, sales, and marketing.

He received a Bachelor of Science in mechanical engineering from North Carolina State University. In 2014, he was selected as a member of the North Carolina State MAE (Mechanical & Aerospace Engineering) Hall of Fame, and he was inducted to the Tampa Bay Business Hall of Fame by the Florida Council on Economic Education. In 2011 he was honored with the Boy Scouts of America's Golf Ridge Council Distinguished Citizen Award, and in 2010 and 2011 Chuck was voted Tampa Bay's Top Business Leader in the *St. Petersburg Times* Annual Leadership Survey of the Business Community. Chuck has a strong interest in economic and community development, education, philanthropy and volunteerism, and sports.

Sykes Enterprises (NASDAQ:SYKE) is a global leader in providing customer contact management solutions and services in the business process outsourcing (BPO) arena. Sykes provides an array of sophisticated customer contact management solutions to Fortune 1000 companies around the world, primarily in the communications, financial services, healthcare, technology, and transportation and leisure industries. Sykes specializes in providing flexible, high-quality customer support outsourcing solutions with an emphasis on inbound technical support and customer service. Headquartered in Tampa, Florida, with customer contact management centers throughout the world, Sykes provides its services through multiple communication channels, encompassing phone, e-mail, web, and chat. Utilizing its integrated onshore/offshore global delivery model, Sykes serves its clients through two geographic operating segments: the Americas (United States, Canada, Latin America, and Asia Pacific) and EMEA (Europe, Middle East, and Africa). Sykes also provides various enterprise support services in the Americas and fulfillment services in EMEA, which include multi-lingual sales order processing, payment processing, inventory control, product delivery, and product returns handling.

Key Theme: Look at the Business Model Considering Digital

"When people are thinking about digital transformation, there's a lot of debate about what that means. I think at its core, it's basically thinking about your business and how you can use today's digital technologies to transform your business from being a physical asset-heavy company to an information-rich company. That's probably it in a nutshell—if every company just went through their business and tried to tackle that theme, I think they would find immense change inside the organization."

Digital Strategy and Company Vision

Tied to this theme is Chuck's strategy and vision for the organization. When he thinks about Sykes' future, Chuck says it is all about "how digital is going to transform the company" in what he terms "the idea economy . . . Because of all the digital accelerating technologies that are out there, we now are becoming a provider of digital customer engagement services and solutions across the entire commerce value chain, not just tech support or post-sale support, and we're delivering it for all enterprises, large and small, as well as consumers."

> I think at its core, it's basically thinking about your business and how can you use today's digital technologies to transform your business from being a physical asset-heavy company to an information-rich company.

This change in business model required building digital capabilities because so much of the existing operational platform was based on large fixed physical assets, which made it cost-prohibitive to serve smaller companies.

Drive Change Inside

Chuck's focus over the last several years has been more on what he calls "the top part of our business model, that's changing our services and changing the target markets that we go after. Now that our business model innovation is behind us, we're ready to start working internally with changing the operation model and doing it in a more informed way because we now have the picture."

Over the years, Chuck has emphasized seven elements for the operational value chain: *talent acquisition, talent development, interaction management, performance management, risk management, workforce management,* and *continuous improvement.* These basically capture the most critical processes that happen in his business.

As the company was going through an immense amount of change, Chuck hired change management experts and trained everybody in the company who is a director and above on change management, preparing them to begin learning the basic methodologies of leading change using the ADKAR methodology (1). ADKAR is an acronym that represents the five tangible and concrete outcomes that people need to achieve for lasting change: Awareness, Desire, Knowledge, Ability, and Reinforcement.

"This is where leadership, integrity, and trust become so important. You have to give people the *desire* when they want the change. And you must also create *dissatisfaction* with where they are today. You have to paint a picture of an exciting vision that people want to be part of—and then you have to provide them a path to get there."

Demonstrating Intent

Chuck believes that integrity and intent are foundational to any aspect of leadership, and particularly if you are having to lead change. "We have a tendency [to believe] that our actions are what they are, but our actions are conveying our attitudes. It's interesting because the truth is, I lead it through my leadership team; if you hire leaders who share your core values, then you are, in essence, establishing a leadership extension of yourself that continues to perpetuate. Over the years, one of the things that I've always shared first and foremost with people is yes, I am the CEO, and that conveys a certain amount of responsibility that I have, but I am not the only leader in the company. I am a leader among leaders, and I always tell my leaders, don't ever lose that perspective."

Chuck says there are three things he expects all leaders to be able to do and compares it to the three legs of a stool. First is delivering results. Second is possessing the ability to build teams. And third is the ability to be a change agent, as that is at the heart of digital transformation.

"Because what you don't know is that even though you may look out on your workforce, you have leaders all among you. They may be leaders in their homes. They may be leaders in their churches. They may be leaders in their communities. And if given the opportunity, you'll see how they possess those leadership qualities. You should always address people with that perspective in mind, that no matter where they are in the hierarchy of a company, you should always treat them with that type of respect. I think it's basic—that has shaped the feeling of our company, and along that way, when you treat people like leaders, in essence, you create the environment where they feel that they should be able to engage with you like a leader. . . So if you come in and you're talking about digital transformation, they're going to have a bunch of questions about how justified is it, or the specifics of what your plan entails that will probably be contestable in their eyes, or they'll have different views about it. And you need to make sure you're not a leader creating an environment of group-think. You don't want to be sitting around the table with everybody agreeing with everything you're saying just because you're the leader. Trust is like a spider building a web; it takes thousands of acts to build that type of trust. You can screw things up pretty quickly, like hitting the spider web with a stick and demolishing it if you really mess up bad.

"Communication around company transformation has to come from a trusted CEO so people can believe you're going to make the right investments to make it happen; they're going to trust in you that they have concerns, and you're going to listen to those concerns because they want to contribute or be part of it. It's a foundational element. If you, as the leader, do not possess the integrity or have the trust of the team or people, and they don't take you to be one who fulfills your commitments and big promises—you're going nowhere."

Building Digital Capabilities Through Acquisitions and Partnerships

Sykes has built up digital capabilities through acquisitions and technology partnerships to really transform their business. Now that the business model innovation has been completed, the company has started to focus internally, to operationalize the way they deliver services. With a greater focus on strategic alignment over the past few years, Sykes has made strategic acquisitions including Qelp (which focuses on the customer journey and digital self-service), Clearlink (which focuses on digital marketing around branded and category search), XSELL Technologies (an agent-assisted artificial intelligence and machine language), and Symphony (which provides a robotic process automation capability). These have positioned Sykes to go to market with a full "lifecycle" platform of capabilities that delivers intelligent customer experiences across the entire customer journey, from marketing, sales, and service to agent augmentation and intelligent automation.

"We've done it across the board—forming partnerships, making acquisitions, and organically doing it ourselves. I think every company, based on their financial ability, should consider those same things. I don't think it's one or the other; we needed to consider all of it."

Operationalization Through Cloud, Data, Automation, and People

Chuck focuses on four key elements to deliver products and services in the newly innovated business model:

1. *Cloud,*
2. *Data-driven platform,*
3. *Automation, and*
4. *People in a flexible work environment.*

First was moving physical data centers to the cloud, which has been a priority. Chuck emphasizes that everything in your business should be measurable, and everything should be known in a data-driven platform. The third element is automation: "There were so many tasks that we do in our business that we believe we can just automate." Finally, people—a flexible workforce, including virtual workforce and micro sites—are critical to successfully deliver products and services.

Results

"We are in a position where we can change our expectations of growth and profitability. We think we can now grow the company 10% a year, on a base of $1.7 billion and our operating margins. Instead of running at 7%, we can also run 10%. We're calling it the '10 by 10' initiative. We are beginning to feel like our messaging to our clients, with our new service capabilities and the messaging about the way that we're changing our internal operations, has allowed us to grow faster, both by capturing new deals and in our ability to ramp up that business."

Chuck thinks that his company is about 30% of the way into their transformation journey. "We're now starting to get some exciting feedback. I think we now have mature capabilities; we have certainly not gotten those capabilities deployed fully the way we want to, but I'm very excited about the future."

Advice to Leaders: Go Back to the Top of the Mountain

Chuck is aware of the immense pressure organizations are under to digitally transform and do something to adapt—and he has felt it personally. He advises leaders to first start on the outside—go back to the top of the mountain, check and scan the marketplace, and assess whether the business model needs to change.

> "Don't start today focusing on how you can better deliver what you currently do. Step back and look at the whole business model and ask, what do we sell, and who do we sell it to? Do we need to change that?"

"When we talk about innovating a business model, there are four key things: What do you sell? Who do you sell it to? How do you deliver it, and how did you make money? Don't start today focusing on how you can better deliver what you currently do; step back and look at the whole business model and ask, well, what do we sell and who do we sell it to?

Do we need to change that?"

Chapter References

1. Hiatt, J. (2006). *ADKAR: A model for change in business, government, and our community*. Prosci Learning Center Publications.

CHAPTER 5.2

Shape Digital Proactively!

ANDERA GADEIB
CEO, DIALEGO

Profile

Andera is an online enthusiast and passionate entrepreneur. She founded Dialego in 1999 with the goal of digitizing professional market research. The founding idea is still relevant for Dialego today, transforming a traditionally people- and paper-intense process of market research into a highly automated business by using technology. Today Dialego is an international market research company with offices in Aachen, Hamburg, London, Paris, and New York. Through customer focus sessions, user panels, global benchmarks, and artificial intelligence, Dialego manages to analyze and understand market segments for its customers.

In 2012, Andera founded her second startup, SmartMunk, a software company for cloud-based customer profiling and market intelligence. She also participates in committees at the state level to advise Germany's political decision-makers on topics related to digitization. Andera studied business informatics at RWTH Aachen University, international business administration at

Maastricht University, and computational statistics and virtual reality at George Mason University.

Key Theme: Shape Digital Proactively

Andera expects leaders to be able to actively shape the digital world. "It's a matter of attitude, more attitude than skill. A leader must feel empowered to shape the digital world. In a digital economy, a leader does not look fearfully from the bottom up at digitization and say that it is all far too complex and incomprehensible. Similarly, a leader in digital transformation cannot be hands-off and only delegate. Shaping digitization proactively means that as a leader, you have to go into the middle of the action and really understand the effects of your decisions and investments yourself—of course, without having to implement them all alone."

Andera believes that "shaping digital proactively" is highly dependent on the context in which companies exist. "It does not mean that you always have to apply the latest technologies without reflection. It does, however, mean that we have to understand the latest trends, take inspiration from what we read, see, and hear, and then think about ways to use it in the context of our companies. That means advanced education, self-learning, and above all talking to each other within the team." Andera promotes this mindset at Dialego through *Dialego Insight*, an in-house program for knowledge sharing where external guest speakers are often invited.

Digital Strategy and Company Vision

Digitization is the core of Andera's business. As one of the pioneers in the market research segment in Germany, Dialego's goal was to digitalize the traditional paper questionnaires and face-to-face interviews. "As a business informatics specialist, I see myself as a bridge builder, a translator from the physical to the digital. In market research, Dialego has digitized all the repetitive tasks."

Andera has also ventured into solving other sophisticated problems, such as evaluating emotions in speech and text. "Discovering and understanding emotions has so far been a human capability, but in the future, it will be done using AI." Her underlying business strategy is to become better at performing repetitive human tasks in market research with the use of technology.

A Digital Mindset Is Nurtured by Courage and Constant Learning

For Andera, the most important quality of a digital leader is having enough courage to go forward, with the necessary power, to achieve their vision. "Digital transformation means an incredible amount of change, an incredible amount of handling the previously unknown. I believe that for everyone, especially leaders—who bear even more responsibility, not only for themselves but also for others—this means moving out of the comfort zone. And, as we all know, getting out of one's comfort zone takes a lot of courage. This type of courage is still very often lacking at the leadership level in established companies and institutions of all types."

Andera believes in experimentation with necessary safeguards to check whether your direction is the right one and to turn if necessary. "The future of leadership is characterized by allowing experimentation and trying different options. At the same time, it is even more important that leaders have the skill to reflect and critique their decisions from an outside perspective and then quickly adapt as necessary."

Andera believes that the attitudes and mindsets of leaders are the most critical predictors of digital transformation success. "In my view, the future we are entering, that we are shaping together, is one in which both leaders and employees are in a position to acquire new knowledge and to engage in new activities. I have to take the attitude that I can do it. That I can make it with my team, that I am not the one who has to know everything, but that I can trust my team. I believe the strongest leader is the one who hires people who can do more than the leader can do by himself or herself."

The CDO Needs Comprehensive Authority

Andera is convinced that a successful transformation must be led from the top down. "I know too many companies that have hired a CDO in the hope that this investment will make the digital transformation a success. Unfortunately, this has failed all too often, simply because a CDO is defined as a staff position with no relevant decision-making power. The CDO can only support but never really decide, and therefore they never actually shape the digital transformation. To be successful, the CDO will have to be a member of the senior management team or on the board of directors and will take ownership of digitization and push it forward consistently and responsibly. In my opinion, this is the most promising approach."

Digital Leadership Means Engaging in Education

As a mother of three children, Andera is concerned when it comes to the education of her children. "When I looked into the textbooks of my 14-year-old six years ago, I saw exactly the same things in computer science that I had learned 30 years earlier. This scared me very much, and when I spoke to the representatives of the Ministry of Education in Berlin, they told me that nobody had to worry about falling behind in the field of computer science. Today, six years later, unfortunately, exactly the same things are still in the computer science textbooks as 36 years ago. Nothing has changed."

Andera emphasizes how important it is to teach the subject of computer science in early school classes and to introduce students to the development of software in a playful way. This is an essential prerequisite for finding the right employees for future-oriented companies and their digital business models. She herself is committed to working with local schools to help children build their first computer and create their own code.

Results

Dialego is a standout performer in the field of digitally based market research, thriving on its founding idea of digitizing human-based processes. This is largely the result of the leadership of its founder and CEO, who spearheads trust-based principles of courage, commitment, personnel development, and constant learning. As Andera puts it, "A leader in digital transformation must always put themselves at the center of the transformation and deal with all facets of digitization facing the organization. They must make a constant effort to develop their employees and to promote a constant dialog within the company."

Advice to the Leaders: Get the Right People on Board

"It's all about choosing the right people for the company. It is even more important to bring in new experience and knowledge from outside. It's not just finding the people with the best technical skills. New employees must have a high affinity with what the company does, and in particular with the topics of digitization and its implementation. You should take the opportunity to find employees who are better than you are at what they do. This is the only way for a company to actually grow beyond itself."

> "You should take the opportunity to find employees who are better than you are at what they do. This is the only way for a company to actually grow beyond itself."

CHAPTER 5.3

Fueling a Startup
Culture and Entrepreneurial Mindset

LARRY BLUE
CEO, Bell & Howell

Profile

Since joining Bell & Howell in 2014, Larry has evolved the company from a production mail equipment manufacturer into a premier provider of industrial automation services. Larry graduated from Duke University with BS and MS in electrical engineering and has over 35 years of experience in product development, sales, and manufacturing. He has a strong record of innovation and success in managing international high-tech manufacturing and product development efforts at companies like Flextronics, RF Technologies, Hughes Network Systems, Hi-G-Tek, and IBM.

A prominent, results-oriented visionary, Larry is passionate about cultivating an environment of change. He instituted innovation labs, championed innovation management processes, and spearheaded the digital transformation that powered Bell & Howell into a technology-enabled, "services first" company.

Bell & Howell is a North America-based services organization and former manufacturer of cameras, lenses, and motion picture machinery, founded in 1907. It is headquartered in Research Triangle Park, North Carolina. Bell & Howell provides services related to automated equipment by leveraging innovative technologies and service capabilities. The company is rapidly undergoing digital transformation and is investing heavily in the tools, technology, and training needed to help its customers increase efficiency, reduce costs, and improve their customers' experience. The company also offers a complete portfolio of comprehensive automation solutions in retail "click-and-collect" and production mail automation. Bell & Howell has over 850 highly skilled field technicians and 24/7/365 customer service and technical support centers, as well as remote monitoring, diagnostics, and advanced analytics capabilities.

Key Theme: Fueling a Startup Culture

The transformation of Bell & Howell, a company founded in 1907, is an amazing success story. Many businesses face disruptive change, and their survival depends on how they traverse the so-called "valley of death" and successfully implement innovation from ideation to commercialization. (1) Studies have shown that one out of ten innovative ideas fail because innovation leaders are often ill-equipped for the journey. A structured industrial innovation process, powered with strong digital leadership, propels a leader to successfully carry out the transformation. Bell & Howell's success, in part, is attributed to its CEO promoting an entrepreneurial mindset so that a "startup" culture can *co-exist and thrive* within an existing legacy culture.

Digital Strategy and Company Vision

Larry led the successful transformation of Bell & Howell from a hardware systems business to a high-tech field services business. His vision is to transform the company into a technology- driven, data-powered, services-first company that drives efficiency and value. Larry sees digital strategy as an integral part of achieving this vision. He foresees that real-time remote monitoring and repair capabilities, supported by analytics, machine learning, and predictive dispatch, give the company the ability to increase efficiency without linearly adding head-count. He has invested in digitization (a field service app, analytics-driven predictive and prescriptive recommendations, live assistance, etc.) to empower field engineers to exchange information and act in real time to enhance the operations and value of Bell & Howell and its customers. Larry believes that digital tools such as the Internet of Things (IOT) and Augmented Reality (AR) will enable "people-based" service business to drive super linear economic growth through increased productivity and entirely new kinds of data and analytics services.

Walking the Talk

Digital transformation is hard; it is a very broad topic, difficult to get aligned across an organization. Sometimes you don't have a culture that is ready to embrace it; you may not know what customers want. Leading Bell & Howell through digital transformation, Larry lives this day to day and is fully aware of this challenge.

"As an executive, people respect what you inspect; if you can't spend personal time really driving digital transformation for your organization, don't expect a staff function or a committee to do it for you. You need to pay attention to it and show your commitment. If you pay attention to it, your people will pay attention to it."

Demonstrating Intent

Larry established a center of excellence innovation lab for advanced analytics, machine learning, and IoT and continues to spearhead investments in people, tools, and technology for digital transformation. Every new product idea at Bell & Howell goes through a proven innovation management process that Larry championed. The simplified process has built-in gates and timelines for each step. The common theme is agility, that is, applying AGILE development concepts across the organization, not just in engineering. Larry stresses fail-fast culture: "Fail fast by tackling the most difficult problem first. Embrace data; data is at the foundation of everything."

Addressing the skill gap in the digital realm, Larry asks his senior management team to "hire a talent and listen to them; make an effort to read; go to seminars on digital transformation; lead by example—get the background, get the knowledge needed to make a decision to drive investments."

Fostering a Successful Culture of Innovation

As a leader, it is important to promote a culture that encourages employees to learn from mistakes. "It is not about getting it right the first time; it's about how well we learn from our mistakes. If you respect the individual and create an environment where creative, data-driven exploration is nurtured, you'll be able to build an innovative culture. Talk about innovation and constant innovation, talk about why it is important; talk about the data, information, unique informational discoveries. Constantly innovate and make decisions fast.

> "Success breeds success. People will see what you are paying attention to; they will want to be part of a successful team, and that will allow the culture to naturally change."

"Start small and build from there. Pick one project to move a digital approach to the business; don't try changing everything all at once. Success breeds success. People will see what you are paying attention to; they will want to be part of a successful team, and that will allow the culture to naturally change.

"Starting small allows you to move fast. There are too many stakeholders in a legacy business, and if you try to tackle everything, you are kidding yourself. You can move fast if you start small. You are a small team, you are focused, and you can make those decisions quickly because you are not dealing with broader issues around legacy businesses and processes. You don't have to take the risk of rushing, and as a leader, you need to be patient enough to take one step at a time. Projects must be nurtured and protected during early stages, and management's commitment must be resolute and visible."

Larry gave an example of one of the early digitization projects at his company to explain how starting small and creating early wins foster adoption. "We look for small successes. We took the data on one product line and started to use it

> "Projects must be nurtured and protected during early stages, and management's commitment must be resolute and visible."

in a brand new way. We were transparent on what our goals were, and we were able to communicate early successes, what they mean in terms of savings. Soon, employees started to notice what we were able to monitor, debug, and save time on. They began to accept it as they saw immediate benefits—better planning, less travel, less wear and tear. At headquarters, we were able to take that data and plan dispatches better. We were saving money for customers, as well, so they also championed it. So key is to look for small successes, communicate, emphasize success internally, and build early adopters. People will join in on successes and join in on broadening the application of technology."

Results

The company embarked upon a quest to evolve its approach to service from a reactive state to a predictive service model in the fastest and most efficient way possible. Three years into the journey, results have been impressive for both Bell & Howell and the customers it serves.

"We are seeing results—we are remotely repairing over 70% of service calls, meaning we don't have to roll the trucks. Troubleshooting time has been cut by a full hour per service call, and our first-time fix rate is over 92%. The real-time data we are collecting has a lot of value to us internally as it drives service efficiencies and adds to the bottom line. These insights help us find out more about our customers' businesses, and we can provide this extrinsic value to make them more productive in their factory, store, warehouse, etc.

"Digital transformation is not a finish line you are crossing; it's a continuous journey. We are probably about 20% along the journey, and we have connected and are practicing digital transformation on less than 10% of our gear. Our 'Horizon 1' goal is to instrument the rest of the equipment in the field, get remote connectivity and real-time monitoring. Our ultimate goal is to provide real-time prescriptive maintenance and advice to improve the productivity and quality in real time while the machine is running to provide a very high uptime and value to the customer."

About 30% of Bell & Howell's current yearly revenue comes from products developed in the last three years—a true testimony for innovation.

Advice to Leaders: Start with the Goal First

Organizations are under so much pressure to move faster for the investment they are making. However, Larry takes a more pragmatic approach, not tackling too many digital projects at one time but focusing on getting to the end goal. His advice is to understand what your goal is and what you are trying to do with digital transformation.

"At Bell & Howell, we were trying to increase the efficiency of the field service organization, and we needed to do that with technology. We started driving a digital transformation by doing more things with the data we collect, going out and collecting more data, and using analytics to change behavior, processes, tools, and procedures."

Chapter References

1. Mugge, P., & Markham, S.K. (2014). *Traversing the Valley of Death: A practical guide for corporate innovation leaders.*

CHAPTER 5.4

Much More Is Possible than Anybody Thought Before

ROBERT KALLENBERG

Director, Strategy and Organization, Porsche AG

Profile

Robert Kallenberg is Director of Strategy and Organization at Porsche AG. Before taking up this position, he was the Head of Strategic Planning for Volkswagen Group at Volkswagen AG. Prior to that, he served as Vice President, Corporate Strategy at Porsche Automobil Holding SE. Robert has worked at Porsche AG for ten years in various leadership roles such as Head of Strategy & Organizational Development and Head of Company-Wide Porsche Improvement Process. His early career also included a consulting role at the Boston Consulting Group (BCG). Robert completed his PhD in mechanical engineering at RWTH Aachen University.

Porsche AG is a German automobile manufacturer specializing in high-performance sports cars and SUVs. The origin of the company is a design office founded by Ferdinand Porsche in 1931, which was merged into an automobile factory after 1945. The headquarters of Porsche AG is in Stuttgart, and the company is owned by Volkswagen AG, a controlling stake of which is owned by Porsche Automobil Holding SE. Exclusivity and social acceptance, innovation and tradition, performance and suitability for everyday use, design, and functionality are the brand values of Porsche. In 2019, the company delivered over 280,800 vehicles to customers worldwide.

Key Theme: Much More Is Possible than Anybody Thought Before

As we conducted this interview during the COVID crisis, Robert reflected on how COVID has impacted the work dynamics in his department. Despite all the demanding adjustments, he sums up the experience from the perspective of digitalization in a positive light. "We did large-scale experiments, virtually, using digital collaboration tools. We quickly moved and set up projects online, although there were brief moments of questioning whether we should stop these projects. But it was obvious that we had to at least try it. Basically, all these trials worked well. Actually, the overall experience has been that it worked exceptionally well. This big shift would not have happened without COVID— and it provided proof of what is possible. Our experience is that much more is possible than anybody would have thought before. Going forward, there will be a new mix between physical and digital. It is likely that hybrid working models will emerge, especially for digital collaboration, and these will exceed our current vision."

Digital Strategy and Company Vision

Robert explains that the digital strategy is an integral part of Volkswagen's corporate strategy. "It is actually seen as business-critical." He further observes that digitalization, from the perspective of Volkswagen, takes place in three main areas: "First, the digitalization within the car. It focuses on connectivity, driver assistance systems, and autonomous driving. In the past, cars were not really digital in any way, but that is changing fast. Cars are becoming part of the Internet of Things (IoT)." Second is "the digitalization along our business processes, in R&D, purchasing, production, sales, and administration." The third aspect of the digitalization strategy ties into the sales processes by "the increasing digitalization of our customers' journey. In the past, this journey was a physical process. People would look at cars using print media and then maybe look at a car in a showroom, opt for a test drive, and so forth. In the future, most of this is happening in the digital realm, sometimes even eliminating steps like a test drive."

> "In the past, cars were not really digital in any way, but that is changing fast. Cars are becoming part of the Internet of Things (IoT)."

All Elements of Trust Must Be Set in Action

Robert emphasizes that it requires all four components of trust—Intent, Integrity, Capabilities, and Results—to create a culture of trust. "I think the intentions dimension, which includes communication, is particularly important. It is essential to send clear messages, to make sure that there are no hidden intentions, afterthoughts, or whatever. Moreover, trust is based on a track record; trust is based on capabilities, and trust has to do with walking the talk. If any of these are missing, or if people feel that there's a major issue with any of these four dimensions, trust is very hard to achieve, and it will jeopardize the overall goal."

Management Should Get Hands-On with New, Innovative Methods

Robert emphasizes that the general abilities of the individual play a decisive role in the search for new leaders: "Skills and knowledge are the most important, whether it is methodology like AGILE or digital capabilities. Specific knowledge of methodology is important—for us, it's especially FuSE methodology that people need to know." Function-based Systems Engineering (FuSE) is a design method that is based on systems thinking and uses functional modeling through product planning, conceptual design, and embodiment design. The objective of the method is to formalize and coordinate all activities that occur throughout the design of systems. (1) "In addition, other methods, especially those to develop software, will become more important. And we need people to understand this. This understanding is actually lacking in many functional areas, particularly by people of my generation. It is not something my generation learned at university because these methods simply did not exist only twenty years ago. Furthermore, our leaders should have hands-on experience with these methods, not just by reading a book. We are looking for people who have real work experience with these methods."

Flexible Work Modes to Attract Digital Talents

"If you have the need to recruit employees with digital skills, and you cannot really get them to move to your location, maybe you can try to attract them by providing flexible work modes. Recruiting this kind of talent is a big problem, because digitally skilled people are in high demand. They typically have an urban mindset, and that is a major problem for many companies that are not located in cities like Berlin or Munich. Try to attract these talents by having a digital office in one of these hotspots. We decided to establish hubs in cities where people with the mindset and the skillset we need actually want to live."

Results

While reflecting on the achievements of the Volkswagen Group in terms of digitalization initiatives, Robert believes that the mindset of leaders regarding digitalization has changed significantly over the last few years. "Five years ago, people were not really aware of the importance of digitalization in our industry. I think that now the mindset has changed. Our newly established car software group is now an integral part of our organization. However, from the outside, there is a long delay in perceiving this due to the time delay between starting the development of new products and launching them into the market. But increasingly people will understand that the digitalization of the car is important and that software capabilities are key competencies for car makers going forward. And our leaders need to firmly believe in it, because we have to put resources behind it, and we need to be able to attract the right talent to drive these efforts. Many of our best people should have the desire to work in this new field. Our new mindset is the key for moving forward."

Advice to the Leaders: Deliver Results

"You need role models and people who have actual experience in digital. Maybe hire some people from outside your own industry if the industry is not digital yet. In addition to digital awareness, it is important to get actual projects done. Make sure to do smaller scale projects first. In the end, it is essential for digital leaders not just to talk about digitalization, but also to prove they are able to deliver results."

Chapter References

1. Hutcheson, R.S., McAdams, D., Stone, R., & Tumer, I. (2007). Function-Based Systems Engineering (FuSE). Guidelines for a Decision Support Method Adapted to NPD Processes.

CHAPTER 5.5

The Trials of Launching a New Digital Business for Family Farmers

BRANDON BATTEN
Owner and Operator, Flying Farmer LLC

Profile

Brandon Batten formed Flying Farmer LLC in 2017. Brandon is a graduate of NC State University, where he earned his bachelor's and master's degrees in biological and agricultural engineering. Brandon started using drone technology after receiving an NC AgVentures grant from NC State and the North Carolina Tobacco Trust Fund Commission to purchase a drone for his family farm. After seeing the benefits and value that the drone could add to their operation, Brandon started Flying Farmer LLC to make this technology available to other farmers. Being a farmer himself, Brandon understands every facet of crop production and can relate to other farmers and their operations, and he helps interpret the data that the drone provides. In addition to operating Flying Farmer LLC, Brandon farms with his father and uncle at Triple B Farms, Inc. in southern Johnston County. They produce flue-cured tobacco, soybeans, small grains, corn, hay, and cattle.

In 2019, Brandon and his wife, Jessica, were selected as the 2019 National Outstanding Young Farmers of the Year at the 63rd annual National Outstanding Young Farmers Awards Congress. The winners were selected from a group of ten finalists for the award based on their progress in an agricultural career, extent of soil and water conservation practices, and contributions to the well-being of the community, state, and nation.

Key Theme: Providing Farmers with Something that Helps Them Still Needs Trust

Starting a new digital business, even in the industry you were born into, takes all forms of trust. By their very nature, family farmers are a special breed. Every year they deal with numerous external pressures—including weather, pests, water (or lack of it), politically motivated trade policies, and government regulations— that would cause most other company leaders to throw in the towel. Yet farmers don't quit in the face of these pressures because they have a particular attachment to their land, and they aspire to a much higher goal—making sure the land can forever produce the products that will feed this nation.

To convince these people to invest their hard-earned profits in a whole new way of managing their land using drones and digital imagery can be a tough sell. Nevertheless, that is what entrepreneur Brandon Batten intends to do.

How Flying Farmer Got Started

Flying Farmer began with Brandon's fascination with technology. As he says, "I've always been a 'technology aficionado.' I guess that came with my engineering schooling; I've always been interested in technology and the *efficiency* that can be realized from its application." His interests meshed perfectly with the explosion of applications for drones and aerial imagery. As he told us, he could "see that this bubble was exploding, and agriculture was an obvious fit.

Look at all the pretty pictures taken of farm fields, and they've got all these color variations, but what do you do with it? And that's when I said, I think I can figure out a way to make this into a real business—without having to hire a whole team to figure out what to do with all the data." With this insight and his passion for technology, Flying Farmer was born.

Digital Strategy and Company Vision

Brandon's strategy, simply put, is to "help farmers stay in business. I want to be able to help them improve their efficiency or reduce use of fertilizer, if they're overusing or overapplying it. I want to be able to help them optimize their systems. Ideally, with what they have, without having to invest in other capital. My vision is to be able to provide actionable solutions to those in agriculture and with agricultural interests, using drone technology and aerial imagery, in whatever form that may take, whether it's just standard pictures, infrared, thermal, or whatever medium they need to accomplish their goal.

"I want to be able to use my unique skill set to provide the data in a way that they can use it. Agriculture, like many other industries, can produce way more data in a short amount of time than most producers are able to use, or even visualize. Being able to put that data to work, I think, is where the challenge is, and that's the gap I'm trying to fill."

Building a Business Based on Credibility

"I would say generally, as we talk about the average farmer right now, they are probably not quite ready for this technology. There's a lot of interest—I get a lot of phone calls, and it's still cool when it's the newest thing, but they're not quite ready to take that step yet. So, I'm hoping to be able to help them be ready to take that step."

One example Brandon gave us was "variable rate technology," where fertilizers are applied dynamically based on the needs of the soil or the crop that is growing and determined by an aerial field map. But as Brandon points out, purchasing the technology requires capital that many farmers do not have, particularly in these times. As he puts it, "I want to tailor a solution built for what that farmer has on his individual farm with what he is trying to accomplish.

"A lot of stuff that comes out in the agriculture space from drones is on an academic level. It's either academic research that's been done or research by the big seed companies. They've done their own research, and it's oftentimes at such a high level that it's not applicable. They're trying to use the latest technology to get the most precise information, but the average farmer doesn't know—or need—that precision to decide what he should do. For example, a lot of people ask me about near-infrared technology. Yes, it's a great technology, but I think you can just use a standard camera and get close enough for 'good enough' information to make decisions without having to spend that extra money."

Practicality and *affordability* are big part of Brandon's strategy—and mission—for Flying Farmer.

Agriculture Is a Handshake Business

Brendan says, "Agriculture has traditionally been a 'handshake business.' People have bought and sold farms and goods for decades and generations with a handshake—and that can be tough. If you're not part of the group, it's a tough group to get into and gain their trust and respect. That speaks tons about the trust that's developed over years. There is no other way to work with farmers—you have to earn it."

Trust is everything in the farming industry. And while being a third-generation family farmer gives Brandon a head start at convincing what he terms a 'tough group' on the benefits of digital imagery, he is not resting on these laurels. He is creating ROI cases for every crop and produce application he has flown.

> "People have bought and sold farms and goods for decades and generations with a handshake—and that can be tough. If you're not part of the group, it's a tough group to get into and gain their trust and respect."

"Creating an ROI takes time because I get only one chance a year, one crop a year to collect that data. So now I've got three years' worth of data; I can put together various use cases and show this is how I've used this in soybeans, in produce, in tobacco, and other crops, and I can show that information. So now that I've got enough data, instead of saying, *This is what I think we can do*, I can say, *This is what I have done*. And that is a very different statement when you're talking to another farmer."

Capturing Results Helps Build Confidence

"We grow winter wheat on our farm that we plant in the fall and harvest the next summer, and you have to apply a fair amount of nitrogen fertilizer. So I did an analysis on one of our fields and took the information the drone provided. With the help of my Cooperative Extension agent, we took tissue samples and found parts of the field that corresponded with various degrees of stress—in other words, the red plants that were stressed versus the lush green plants that looked healthy. We found that the areas where the drone showed the stressed plants—through the tissue analysis in the soil samples—needed about 30 more pounds of nitrogen fertilizer to produce their potential, while the other areas of the field that were already green and lush on the drone imagery didn't need any additional nitrogen. The traditional way to fertilize is to spray the whole field with more nitrogen to make sure it is all covered. But using variable rate technology, you could reduce the amount of nitrogen needed on just that one 15-acre field by 63%.

"For farmers who find variable-rate application equipment too expensive, I have devised a new method. I still fly their field, but now we divide the field into 'strips' the width of their current fertilizer equipment. Now the farmer sets the fertilizer application rate for that strip. We don't get the same precision we would have using variable-rate technology, but we can still save up to 30% of fertilizer! Putting this extremely technical information and data in language that they can understand and that they can use is still my toughest challenge."

Walking the Talk also Builds Credibility

Flying Farms' target market is family-owned and managed farms, not large corporate farms. We asked Brandon, why do you think the family farms are more important that you need to help them to survive versus corporates? Brandon's now obvious answer was, "I'm a family farm. I live on a family farm.

> "I'm a farmer. I do this, too, and I'm using this technology. This is how it's helped me; this is how I think I can help you. And that increases my credibility."

"You get a lot of benefits, like opportunities of scale when you are a large farm. You can buy more inputs, cheaper, etc., and you can also afford to hire somebody who can do this for you. Whereas in smaller farms, where the average age of the farmer is, I think, 60 years old, *they're just not comfortable with this technology.*

"And if you were to send out a masters or a PhD from a university and say, 'Hey, I'm from the government, and I'm here to fly your farm,' they would probably run you off the property. I have a farming background. I can talk the talk, if you will. I understand what it is to farm every day, and that helps me build trust with those potential customers. I'm a farmer. I do this, too, and I'm using this technology. This is how it's helped me; this is how I think I can help you. And that increases my credibility."

Educating Users is Vital

"I give a lot of talks. I go to a lot of farm meetings and groups that are interested in drones. Just this year I talked to the soybean producers' association, just putting the information out there. I'm using these commodity groups and different groups to get the information out there, to just show people what the capabilities are.

"And I haven't told anyone no. If somebody asked me, do you think we can do this with the drone? If I've never done it before, I'll tell him I don't know, let's try it and see, and that has opened a lot of doors for me. We've been able to do a lot of things that I wouldn't have thought possible, just because we tried it and it works."

What Is Your Ultimate Goal?

We asked Brandon to tell us about his ultimate goal. How do you make a real change in the world using digital technology in farming? Would you use the technology to drive down the production (and application) of sometimes toxic fertilizers, increase crop production, or address the planet's *biggest* goal—solving world hunger?

"The answer to this question has really changed, especially since I've had children. I am concerned about the environment because the environment— the land and the water—is how I make my living. My ultimate goal is one day, should my children desire to farm, that I can provide that opportunity for them, just like I was given that opportunity.

"There are fewer and fewer people involved in producing our food, and those people are having to do more with less—less land, less water, and regrettably fewer farms. My goal is to help every producer stay in business because I believe that the value and the strength of our food system is in its diversity. We may have a disaster in the southeastern United States, but the Midwest would be well and could fill our gap in production. Or vice versa, there is a disaster in the Midwest, and we can fill their void—even with fewer farms. I have only 24 hours a day just like everyone else. The more I have to do with technology, the more confidence I have in the efficiency of US farms.

"I think drone technology, or any technology, is just another tool in our toolbox that we can use to do more with less. I like to use it as a means to build consumer trust as well. The drone is something that a consumer has seen and can relate to. When I tell them that I'm using it in agriculture, then that opens a door for me to have the conversation about the huge challenges that face us."

Advice to Other Digital Leaders: "Do it Again (and Again) Until you get it Right"

"Before I launched Flying Farmer, I suffered from 'analysis paralysis.' I could only think of reasons *not* to do this and why it wouldn't work, and nobody's going to hire me to fly a drone! But I founded Flying Farmer anyway. And I think that's my advice to entrepreneurs trying to form a new digital venture: Put aside your fears and do it anyway. You either succeed or you learn; I don't really look at it as failing. Then do it again until you get it right. It's not a linear path, it's iterative. You have to do it and then do it again and keep doing it until you get where you want to be—or until you are confronted with the next challenge you feel compelled to solve."

Special Message for Young Entrepreneurs in the Farming Industry

"I've met a lot of people who are interested in farming or who think they want to farm. Maybe they don't come from a farming family and don't understand that farming is so capital-intensive. It is really difficult to just start farming from scratch, but there are other ways to work in agriculture. And this is one way I talk with student groups all the time, telling them about *technology*. if you want a career in agriculture, there is a place for you, and it's really eye-opening to tell them that. It gives them confidence to pursue that career in agriculture. Agriculture is North Carolina's number one industry. We need the best and the brightest for those reasons.

"As an engineer, I was always trying to solve problems in school. And I think this problem of feeding the world is probably the biggest challenge that we're going to face, at least for the foreseeable future. I'm the third generation on our operation, but one of the most dangerous things is people saying, 'That's the way we've always done it' and being unwilling to change or try something new. Luckily, my father, uncle, and grandfather were very open to new ideas and trying new things. Especially my grandfather—he was always very aware of the latest technologies and thinking of how we could apply that here on our farm.

"'Let's NOT just keep doing the same thing' is gaining a lot of traction with the younger generation. It's giving them the confidence to have that conversation with their dads and granddads and saying, 'This is a new technology, and I think it can help us.' And, of course, backing this up with real examples (back to the ROIs) of the services Flying Farmer offers is a relatable story to my customers and to these young entrepreneurs."

Brandon adds, "The main thing when you're educating people about agriculture and sustainability is that *farmers are environmentally conscious.* Technology is changing so fast, and everything is getting smaller and cheaper. I don't know what the future holds. But I'm sure that I—or other young digital leaders—will be able to find the novel use for it in agriculture, ensuring that the United States still has the most sustainable and affordable food supply and that we are able to provide these products to the rest of the world as well."

CHAPTER 5.6

Transformation "Evangelists" Are Vital!

MARC SCHLICHTNER

Principal Key Expert and Founding Member of the T-Club, Siemens Healthineers

Profile

Marc works in portfolio management at the Digital Health unit at Siemens Healthineers. He also founded the T-Club, a transformation community promoting interdisciplinary exchanges and digital transformation. Before joining Healthineers, Marc held various leadership positions at Siemens AG, such as Head of Technology and Innovation at the Service Unit of the Digital Industry Division, Head of Portfolio Management Digital Services or Head of Group Modernization and Migration Solutions. Due to his many years of experience in product, portfolio, and innovation management, he holds the function of a Principal Key Expert at Siemens Healthineers in the respective fields.

Siemens Healthineers is one of the world's leading medical technology companies with over 120 years of experience and 18,500 patents. The company helps healthcare providers worldwide to expand precision medicine, transform healthcare, improve patient experience, and digitize healthcare. The portfolio includes ultrasound examination systems, mammography devices, computer tomographs, and much more.

Key Theme: Transformation "Evangelists" Are Vital

Although Marc has a classical line role within the company, he is also a lateral thinker and strives to transition towards a more innovative and explorative culture within and outside the company. For him, this mission as communicator and translator is particularly important. "Employees and leaders should be able to translate the transformation sermon. It's not about technical or psychological skills that we're lacking, it's about bringing them to action. What is the missing link? Actually, it's really intertwined with the evangelist and the topic of transformation, because everybody involved in such a transformation—just from a hierarchical, organizational standpoint—needs to have the same mental model and mindset."

Digital Strategy and Company Vision

During his career, Marc has seen many good projects and setups that did not gain the desired traction because "translation" between stakeholders—and especially decision-makers—did not work sufficiently. Translation processes are at the heart of his transformation sermon: "Basically I'm trying to motivate people who are doing the right things by helping them translate their messages to management and vice versa. At Healthineers, the transformation process happens both 'bottom up' and 'top down,' because it is best when they both take place at the same time. Grassroots revolution—or rather *evolution*—this is the T-Club. It has clear top-down support and commitment by the C-level sponsors; it's not just lip service. It is really intentional. That's when both synergies come together."

Transformation: The Infinite Game with Minimal Personal Reward

In Marc's opinion, a continuous transformation process is an "infinite game," and it is needed by every company to ensure long-lasting success. "If you want to drive digital transformation, you try to get your company fit to play the infinite game, meaning you want to keep your company in the game forever." In general, for leaders, this long-term game is sometimes in conflict with short-term targets and also with promotional elements of a typical career path. "Forever sometimes contradicts making a career; it might mean that you have to do something new that you don't get credit for. You might burn money; you might fail to learn. This is not always accepted, especially in large companies. Hence your intention is crucial. What's your motivation? Because if you are a classical line manager, motivated by exploitation, career, and power, this explorational, experimental mindset is in strong contradiction to everything you have done before. You need to 'disrupt' your own mindset first, and that's not easy. On the other hand, classical line management also has light and shadow to it. Don't underestimate the price of this to you and your family. How many things will be left out of their lives? Seeing your children frequently vs. traveling around the globe. Just imagine how big the personal threat is to those working on transformation initiatives, which may fail if they put their career—where so much personal investment has happened—at risk."

Create a Supporting Transformation Network

Through his many years of professional experience, Marc has witnessed many leaders in transformation projects. In fact, he has held a leadership position in many explorative activities himself. A central conflict during digital transformation, especially in large organizations, is the clash of the two different worlds. "When you're in a large company like ours, you have two worlds. Exploitation conducted by the 'classic' line management and 'exploration,' where you search for future

> The team wants freedom, and the line wants results, and they have no understanding of each other.

business models with a kind of entrepreneurial startup mentality. Leaders of a digital transformation find themselves caught right in the middle, where the established hierarchy is always putting high pressure with exploitation questions and KPIs to the exploration teams, which sometimes almost stabs promising initiatives. So if you find yourself in this challenging reality, it is crucial, as the leader, not only to have the right mindset, but you also need to have a good network so you can survive for a certain period of time at this most challenging and therefore also weakest point in the system." Marc further describes this position as full of ambiguity. "The team wants freedom, and the line wants results, and they have no understanding of each other. If you're trying to do a transformation in a hierarchical system, the leader needs to be very strong. I've never seen one survive more than two years."

Translate Everyday Experience to Your Business

Marc believes we all face new business models and new technologies on a daily basis in our private lives. We are consumers too. For him, a good leader must be able to reflect on these private experiences and apply them to his own business and his forward strategy. "We have events like Black Fridays. We consume subscriptions. We see how we are 'manipulated' every day by companies that are very successful. It's not a matter of not knowing what happens, because we experience it every day. What is really hard for most of the leaders—and also employees—is to translate this to their businesses. I would say it is a very important capability to be able to reflect on what you experience in everyday life and then be able to translate this into your business."

Let Your Blind Spots Be Revealed

Marc is convinced that the knowledge that helped many senior executives reach their positions will not help them to pave the way to a successful future. For him, dealing with these decision-makers is crucial. "First of all, leaders have to be aware of their blind spots. They have to understand that there is a certain pressure in the market to change. And trust me, everybody in the C-level knows that this is a topic. They know that new business models are required, and they believe they have the answers—but many times they don't."

"The first thing they need to understand is that they have blind spots. The questions are 'Who is allowed to tell C-Level managers about their blind spots? And who is allowed to answer C-Level managers' questions about their blind spots?' The natural reaction would be, 'Well, somebody in my direct line that I employed'—the classical line management thinking. But maybe they have to accept that in an unknown system, the classic hierarchical line management knowledge is not the only and the best anymore. A progressive and innovative approach is needed to overcome this systemic hurdle."

Use Incentives Adapted to Personality Types

Since the so-called "digital rock stars" are very hard to find, it is very important for Marc to attract them and keep them in the company. "However, caution must be exercised, because the motivations of these employees can be fundamentally different from the traditional ones. I'm sure there are more than a thousand books written about people's different motivations. And if you say to somebody who is solely motivated by delivering results, innovation, and exploration, 'I am promoting you,' you're actually doing harm. You put them in a position where they don't necessarily want to be, and you are giving them an incentive that these people probably couldn't care less about. In a sense, you're actually punishing them with the best intent. Furthermore, you maybe harming your company because you're putting them in a position where they cannot deliver highest value, and they may even get frustrated. Subsequently, the only thing he or she can do is to leave, which no one really wants.

"Motivate your employees by using incentives they really want. Give them freedom, give them an audience, let them show the results they created, and give them an alternative career path to the classical line management." In other words, guide and shield them. Also, small gestures can be of crucial importance, as Marc reports in an anecdote from the T-Club: "I suggested that the CTO give personal credit to those supporting transformation. 'Here are the first three phone numbers—please just call them and say thank you for being an active T-Club member.' And he did. The next day one of the T-Club members called me and said, 'Do you know what happened? The CTO called me.' And this person was happy for two weeks."

Not Recognizing Technology's Full Business Potential Is a Threat

"Due to the fact that we are quite excellent in technology at Siemens Healthineers, new technologies are not the biggest threat to our company. Enough people are deeply involved in these topics." Nevertheless, Marc urges caution, as new business models, products, and services with immense potential for the future may not be followed up with required focus and energy. "I don't even have to worry about someone taking care of AR (Augmented Realty) because many people are. I don't have to bother about somebody taking care of blockchains because many people are. I don't have to bother about AI because a whole team and a product line are thinking about how we can apply this to, for example, support radiologists. What I'm afraid of is that we don't look at the business models and business potentials, where other companies are applying standard technologies with new services and business models in a much smarter way. They have the potential to disrupt parts of our business.

> "A bad analogue process digitalized is a bad digital process, meaning all you did is automate what was not of value upfront."

"It is always important to understand technology as enabler. A bad analogue process digitalized is a bad digital process, meaning all you did is automate what was not of value upfront. Our job should be to think about what we can do from a value and business perspective, internally as well as externally. This is the focus we need to have in our transformation."

Experimental Culture Dispenses with Safety Nets

To protect yourself from being disrupted, experiments have to be part of the digital culture, which makes it possible to promote innovation quickly and to sort out unsuccessful projects on the journey to search and learn towards your next big thing. However, it is precisely these rapid project process changes that pose a challenge for large, western companies. "I always say there are two hard things for a large company. The first is to start a project, and the second is to stop one. And this is absolutely in contradiction to exploration and required experimentation."

Marc is convinced that corporate culture plays a decisive role and that safety nets are not required. "One of the biggest goals is what we call 'transformational DNA.' As soon as the company culture is there, you don't need the safety nets anymore because nobody will be punished for doing an experiment. If you don't get punished, why do you need a safety net? It's a cultural topic. If you give scared people a safety net, they will ask for more safety nets, because they're lacking the courage."

Results

"The most positive results I saw are when you really stick to something over a long period of time and where you follow your vision, just keep going, step by step towards your goal. It's good to see that even if you leave a job, people still use the same methodology you introduced. Once you leave, and people continue doing what they have learned, this is when you see that there has been value in it for them. If you leave and they stop doing what you introduced, there may have been value, but nobody really saw it."

Marc points out that over his career, those projects that created a fundamental change, even after he left, created the biggest impact. This highlights the fact that results sometimes come after a certain latency period and unfold after leaders are already focusing on new tasks.

Advice to the Leaders: Moving the Limits of What Is Feasible

"Look at what's happening around us, especially in the consumer world, which is probably the fastest moving industry. Just see what could be applied and make some crazy assumptions. For example, you see Alexa. You wouldn't put her into an operating room and have the doctor ask, 'Alexa, please tell me how I can improve my procedure.' But why not? Why can't the doctor say 'Alexa, I have the following issue. Which doctor should I connect with to help me save this patient's life?' What would happen then? And think how quickly this links to natural language processing." Marc further points out that the current pandemic comes hand in hand with a huge potential for innovations likes this. He urges digital leader to "stop thinking in deadlocked patterns."

CHAPTER 5.7

Soft Skills from a Leadership Perspective Can Be the Difference-Maker

SETH KAUFMAN

President & CEO, Moët Hennessy North America

Profile

Seth Kaufman is the President & CEO at Moët Hennessy's North America division of the wines and spirits sector of LVMH, the world-leader and creator of luxury brands that are recognized around the globe.

Prior to Moët Hennessy, Seth was an 18-year PepsiCo veteran, most recently serving as president of US nutrition businesses where he expanded and transformed PepsiCo's multi-billion dollar portfolio of nutrition brands. He's also held the roles of Chief Marketing Officer, PepsiCo North America Beverages (NAB), Senior Vice President, Pepsi Trademark and Flavors, General Manager, North America Coffee Partnership, and various other media and marketing positions within PepsiCo. He has received numerous accolades including honors from International Television and Radio Society's Hall of Mentorship, American Advertising Federation's Hall of Achievement, *Business Insider*'s 50 Most

Innovative CMOs, and AdAge's Media Maven. Seth also serves on the board of the Immune Deficiency Foundation.

Seth received his Bachelor of Science in television, radio & film management from Syracuse University and his MBA from University of Michigan.

Key Theme: "People First" Is My Theme

Digital is an integral part of an overarching business strategy. Success of digital comes down to people and the leadership team's emotional intelligence and soft skills. It's about speed, agility, empowerment, humility, empathy, and trust. "When you start from a place of people first, empowerment automatically becomes part of the leadership model for which trust is essential."

Digital Strategy and Company Vision

Seth saw the emergence of digital firsthand while running consumer engagement and media for Pepsi back in 2009 when the question on everyone's mind was about digital strategy or social media strategy. "It's not about having a social media strategy or digital strategy; it's about having a business and people strategy, of which digital is a huge piece. Any successful organization with a sustainable business strategy should have digital embedded throughout.

"It's eCommerce, its digital solutions, it's digital media, it's a digital forum, digital coupons, mobile payment platform, or virtual New Year's Eve champagne celebration during the COVID era—all this is digital, but it is part of our company strategy. It's more about having the right overarching strategy, and then, in a people-first, empowering way, allowing the team to make digital a bigger part of the total strategy."

Servant Leadership: Trust and Empower People

Seth is a firm believer in servant leadership, that a "leader's job is to set the vision and the strategy and connect the dots. The leader should spend all his or her time, energy, and effort on breaking down barriers and obstacles that get in the way of achieving that vision and strategy." When you start from a place of people first, "empowerment automatically becomes part of the leadership model and part of how the organization operates because without absolutely deferring decision-making, you're not really thinking in a people-first manner.

"But to empower, there has to be trust. And that trust can be built in a number of ways. It could be trust that already exists because you've worked with individuals before, or it could develop over time with trustworthy actions, but you just have to start from a place of trust versus a place of mistrust. And until I'm proven wrong, I trust the team and I always have. That's not to say that if I'm proven wrong, I don't address it. I do, and that's absolutely critical, because if you don't address it, others lose trust in you."

Empower and Learn from Mistakes: Victors from Failure

"An important facet of trust is giving the leeway to make decisions even if those decisions end up being the wrong ones. In my mind, you learn from mistakes as much as you learn from someone telling you what to do."

These are not just empty words. In fact, one of the year-end awards that Seth introduced and is quite passionate about is called Victors from Failure. "It celebrates a giant mess-up in some way, shape, or form—a bad decision, bad execution in the marketplace, something we shouldn't have done. But you're not talking about the failure; you're talking about what you learned from it. If I trusted and empowered this team, if they did 80%, 90% of it well and then 10% wrong, that failure is okay. Otherwise, they're not going to feel empowered, and we'll break down trust, we'll lose trust. It's so much easier to lose trust than it is to build trust. It takes probably 10 trust-building things to make up for one trust-busting thing, whatever that ratio is."

Conversations with Seth: Humility and Empathy

How do you build trust? "For me, it starts with humility as the senior-most leader in the organization as well as the leadership team. If we didn't have humility, and we thought we had the answers, it's hard to walk the talk. We spend a fair amount of time as a leadership team talking about the softer side of things, and then they spend energy with their teams on deploying them."

When Seth came to the organization, he conducted employee round tables called Conversations with Seth. "I got to hear the pressure points of the organization." When the company did a significant reorganization in support of the new strategy, "so much of that was shaped by what I had learned, heard, and shared with my leadership team." Forums like that gave leaders the opportunity to align the organization around a strategy and assess gaps in capabilities. When those changes or new capabilities were launched, "it didn't surprise people, and they felt like they were part of what shaped it. Those conversations were real. We actually were listening. And for me, that starts with humility and having real empathy for employees."

Building Capability Through SLAM Teams

When Seth joined Moët as its CEO, he quickly realized that the company's eCommerce approach of selling wine and spirits directly to consumers was fundamentally broken. "We had a platform where we lost 92 out of every 100 consumers after one experience, and we knew we had to accelerate our eCommerce efforts."

He and his leadership team made a difficult decision to take down the eCommerce platform for a complete reset. He championed a SLAM team, something he used at PepsiCo. "We brought together a group of people across the organization who are passionate and energetic irrespective of reporting structure." SLAM stands for Self-organizing, Lean, Autonomous, and Multi-disciplinary. "When we combine these characteristics, we foster agility, alignment, collaboration, and speed. Despite our size, we strive to act more like a network of small, tightly knit teams. By organizing around the work to be done, rather than the lines and boxes of our matrixed org chart, our teams avoid becoming siloed."

In a sense, it wasn't self-organizing; the members of the team were hand-picked, but the team owned everything else themselves. "They built a blueprint that was not only inspirational, clear, and helped us understand where we needed to put resources but also enabled us to quickly implement elements that had a material impact in the marketplace."

It served the company well; the timing was just right. When the COVID crisis hit, this team was literally in the process of finalizing their blueprint. "We quickly gave them the green light. They went right into implementation, and within a period of two months, not only had we regained all of the share that we had lost, but we also now have much more market share online than we have offline. And we're in a much better place than we were at the very beginning of the crisis.

"We trusted this team. We empowered them. We had check-ins with them. We talked about the barriers and obstacles. And we would not be anywhere close to where we are today performance-wise if we had tried to run it out of the leadership team or if we had tried to run it in any traditional way."

Data-Informed Organization

One of the lessons learned during employee round tables was that "we are not data-informed; we had not fully digitized the organization, and we weren't making data-informed decisions." Seth saw this as a massive opportunity. He hired a new head of Strategy, Data, Consumer Insights, and eCommerce with a goal of building the data-informed organization. That included measuring and eliminating all sorts of waste, simplifying, and streamlining data pipelines and sharing data to make informed decisions.

Culture Finds a Place in Strategy Blueprint: Inclusive, Transparent, Agile, Empowered

Seth firmly believes that a workplace where everyone brings their entire selves to work rather than just part of themselves is a much better culture. At Moët, culture is part of business strategy. The four underpinnings of culture—inclusive, transparent, agile, empowered—feature prominently in the company's single-page 2030 strategy blueprint. "These are the things that we're going to do every single day in our culture."

This message is personal to Anu Rao, Vice President, Communications at Moët Hennessy North America, who joined our interview on Seth's request. She added, "Communications is a key vehicle in building culture, and I hold

> "At the end of the day, being inclusive, being transparent, being empowered, being agile are at center of an organization that has trust."

myself accountable from a KPIs perspective to try to bring people together to get us on the path towards inclusivity, empowerment, transparency, and agility from both work and personal perspectives. We care about our employees, and we want to hear what they're thinking, and we want to talk to them. We want to open those lines of communication."

Seth adds, "At the end of the day, being inclusive, being transparent, being empowered, being agile are at center of an organization that has trust. When you give trust, you become more agile. I just believe strongly that we must give employees permission to not leave part of themselves out and rather get to a place where we have a culture that trusts one another and moves with pace."

Results

Seth's focus and investment in the digital realm of the business have shown significant results as evidenced by the success of the new eCommerce site, where the company regained market share in online sales.

"We're going to double our profit by the year 2030. That's where we start, but within that, there are key digital levers that can get us there. One of our key metrics for eCommerce is to achieve 20% of our off-premise business—that is when wine and spirits are purchased in a store to consume at home—to come from direct-to-consumer through one of our digital platforms."

Advice to Other Digital Leaders: People, Culture, and Soft Skills

Seth underscores the importance of a leadership team's soft skills. "The biggest driver of digital leadership is high EQ, Emotional Intelligence. I know that sounds odd given how technical digital is. But it comes back to speed, agility, empowerment, humility, empathy, and trust. And without leaders having high emotional intelligence, it's just impossible no matter how technically savvy you are, no matter what your vision of the future of digital landscape is, you won't be able to get an entire organization there. My encouragement to other leaders is inarguably in one of the most technical areas of business leadership; the softest of skills from a leadership perspective can be the difference-maker. And I firmly believe that."

> "Inarguably in one of the most technical areas of business leadership, the softest of skills from a leadership perspective can be the difference-maker."

CHAPTER 5.8

Stop Experimenting with AI–Scale It!

DEBORAH LEFF

**Former Global Leader and Industry CTO of
Data Science and AI, IBM**

Profile

Deborah Leff is a leading expert in outcome-driven business transformation. She is a proven advisor to senior executives on successfully identifying, prioritizing, and delivering on strategic AI initiatives that impact their most critical business objectives.

Currently, Deborah is an independent consultant, advisor, board member, and frequently requested public speaker. She served as the Global Leader and Industry CTO of Data Science and AI at IBM. In this role, Deborah worked with senior leaders of Fortune 1000 companies, helping them gain critical insights from data to drive improved customer experiences and optimized business operations. Prior to IBM, Deborah was the SVP of Business Development at GyPSii, a provider of geosocial networking applications and services, and she was the VP of Sales at Kadient, Inc.

Deborah joined the advisory board of Recruiter.com in July 2019 and was subsequently appointed to their board of directors in August 2020. In this role, she focuses on the intersection of human capital management and AI to develop strategies for eliminating bias in employee recruitment and retention, as well as leveraging AI to drive employee growth and development and enhancing the overall employee experiences.

Deborah also founded Girls Who Solve in 2019, an enrichment program for high school girls designed to spark interest in using technology and data science to solve real-world problems. Deborah uses the AI that students engage with every day and takes them beyond the scenes to understand how things work and why. Hands-on exercises teach participants how technology allows us to solve problems, capitalize on opportunities, and impact our world in new ways. After only one year, this program is in high demand, and she has been asked to replicate it in high schools across the country.

Key Theme: Experimentation Is Good—but Full-Scale Production Is Better

Deborah shared with us that over her career, she has worked with all levels of management and been involved in a wide variety of AI projects. In her role at IBM, as CTO for Data and AI, she met with executives around the globe at some of the largest companies in several key industries during the last number of years. In many of those conversations, she observed a recurring theme: Most executives thought they would be further along with their AI initiatives than they were. It seemed that an inordinate number of projects had not been delivering the value they had expected—they either get

> … most executives thought they would be further along with their AI initiatives than they were. It seemed that an inordinate number of projects had not been delivering the value they had expected—they either got stuck in experimentation or took significantly longer to put into production than they had anticipated.

stuck in experimentation or took significantly longer to put into production than they had anticipated.

"This may be hard to believe, but I have seen companies really struggle to put these advanced technologies to work. Most have hired talented and capable data science teams and are very proud of the work that they are doing. The conversation got tougher, however, when I'd ask about their successes and turn attention towards what had successfully been put into production. That's the point in time when it felt like the energy suddenly drained from room. Very few companies I worked with are advancing on their AI agendas as quickly as they'd like."

This observation is consistent with a paper published last year by *Harvard Business Review* titled "Building the AI-Powered Organization." (1) According to the *HBR* article, of the thousands of executives they polled, "Only 8% of firms engage in core practices that support widespread adoption. Most firms have run only ad hoc pilots or are applying AI in just a single business process."

Deborah says, "Given our world is changing rapidly, companies need to be able to *respond and adapt in near real time.* Getting to AI at scale is no longer a nice-to-have, it's a must-have, and executives are feeling the pressure to modernize—especially if they are a legacy company competing against digital natives. It's not enough to infuse intelligence into the organization in small pockets; they need to become a data-driven organization.

"Discussing how you might want to leverage AI is actually the easy part. The hard part is the access to data, limitations of siloed and rigid legacy systems, cultural impacts, and the fact that AI is not additive and cannot be easily added to existing processes or applications. I think a lot of the early press on AI success focused on how magical the results felt and almost made it seem like it was possible to sprinkle 'AI pixie dust' around an organization and make everything more intelligent, but alas, that is very far from reality."

Digital Strategy and Company Vision

So, what's holding companies back? We asked Deborah, based on her experience, what should their strategy be to break out of experimentation and go into full-scale production? She promptly replied that "these companies must: 1) focus on outcomes, 2) involve the right people, and 3) align AI initiatives to strategic objectives."

Focus on Outcomes

"It was very exciting to be a part of IBM when Watson was launched to the market. It was one of the first enterprise-ready Artificial Intelligence solutions available, and it was the dawn of a new era about how human intelligence can be augmented by machines. People were so curious to discuss what AI could do for them. I cannot tell you how many times a client asked for a meeting to learn about Watson, and when I'd ask, 'What about AI do you want to discuss?', they almost seemed confused by the question. AI is a very broad field; it took some time before that was fully understood.

"There was so much fanfare about early successes that AI quickly became a board-level discussion. Suddenly every company needed an AI roadmap. Now successful technology projects have been happening in every corner of every organization for years, but one of the keys to success with AI is to shift the perspective away from *what the technology is* and onto *what it does*.

"For a long time, innovation meant, 'We are on version 11.2 and are looking for enhancement ideas for version 11.3, and we want to go live in two quarters—so be reasonable.' This has trained us to think about innovation in the context of what we have, and that's not true innovation. We must shift the conversation away from data and technology to focus on outcomes in order to break free of historical constraints—and open the door to innovation that is rooted in problem-solving." Deborah wrote an article on this issue on Medium. com titled "AI Demands a New Perspective." (2)

"To bring this to life, I ask people to picture a grocery shopping cart in their mind. Can you see what it looks like? I then show them a picture of the actual patent application drawing that was filed in 1939, and we discuss the innovations of the cart over time. In 80 years, here are the advancements: bigger basket, bigger wheels, and a child seat with safety belt. That's it. We all have to shop for groceries in order to feed ourselves and our families. It's a task that many of us do not particularly enjoy; we view it as a chore, and any store that can elevate the experience is likely to attract and retain a strong customer following. If you were an executive at a large grocer, how would you innovate? In the past, you would study the cart and think, *What can we do to make this better?* That's how we arrived at the bigger baskets, child seats, and wheels that don't go wonky mid-aisle. What AI needs from us is to think about the *shopping experience* instead of the cart itself. For example, customers don't like waiting in line to check out. They get frustrated when they reach the till only to remember an item they neglected to get, and now they have to decide if it's worth the effort to run through the store to grab it or just forget it. It's inconvenient to lift all the groceries they just put into the basket onto the belt, only to be rung up and put back in the basket. If you have allergies or dietary restrictions, you are constantly reading labels, which is an added chore. Now with all that in mind, how would you redesign that cart to make the experience better? This is exactly the idea behind Amazon Go outlets.

"It's not too late to run your project list through this lens! It's where the most successful transformation is coming from. That's how the digital disruptors have entered into spaces and have completely upended markets. Not because they had any different technology than anyone else but because they were thinking about serving the customer in a way nobody else was thinking about.

"There is another reason why this is an important exercise. It ensures that you are fully scoping the project all the way to production, because you are including how users will interact with the models. I think a major reason why so many projects get stuck in experimentation is because that is how they were scoped— the scope was only what it would take to prove that a machine-learning model could deliver new insights. When I read articles about the high rate of machine-learning projects failing, I don't think that it's the data science that is failing. I think the issue is that companies are getting stuck in experimentation because it was only provisioned as an experiment, and they did not define upfront how that insight would be used at the point of impact. Without interest and support to move it all the way though production, the experiment remains just an experiment. Companies should think bigger from the get-go and focus on how they will impact the entire customer experience before they start."

Involve the Right People

Deborah fervently believes that "digital transformation is a team sport." As she puts it, "Digital transformation has to have all the right cross-functional stakeholders engaged, with everyone working towards the same goal to get it done. And when that happens, the things that companies can do are magic.

"Having the right cross-functional team is critical on several levels, but the most important reason is that collaboration will help shape the project. I have seen the IT organization and data science teams join forces to solve very important business problems—only to see their amazing work go nowhere. The business owners think they solved either the wrong problem or the right problem in the wrong way. Worst of all, they did solve the right problem (and in the right way), but the users don't trust the model and simply reject it outright.

"These investments are too important to risk not getting it right and facing the issues needed to adopt AI. Business leaders have critical domain expertise and must have a seat at the table. I would go so far as to advise that business leaders lead any projects directly impacting their P&L responsibility. A few years ago, I read an article that predicted, 'AI will not replace managers, but managers that use AI will replace those who don't.' AI has the ability to augment human intelligence. We live in a very technical world, and problem-solving with the help of data and technology isn't just the future—it's our present. But I bristle every time I hear a business leader shy away from being involved in machine-learning initiatives under the cloak of 'I'm not technical.' Personal growth and career development require that business leaders have a basic understanding

> "When I read articles about companies getting stuck in experimentation, my thoughts are that it's not that the project itself is stuck—it was only provisioned to be an experiment. Companies should think bigger and focus on the entire customer experience."

of AI principles. The greatest investment a company can make right now is making sure their leadership team is comfortable in this arena."

Align to Strategic Objectives

"Every company should have a documented AI Roadmap that starts with the strategic objectives of the company for the next 12 months. Each P&L owner should participate in discussions about existing obstacles that prevent them from achieving their goals and opportunities for them to overachieve. I often ask executives, 'If you had a crystal ball and could ask a question about what is going to happen—in a week, a month, a year—so that you could make a better decision today, what would that be?' Then you vet the candidate projects for the AI Roadmap. You test their feasibility and ensure that they are outcome-driven, fully scoped, have the right stakeholders involved, and have measurable success criteria. All other requests for experimentation should be evaluated against the AI Roadmap to make sure the effort aligns and supports it—otherwise don't do them. You would be surprised at the number of projects I have seen move forward simply because an individual had an idea they were passionate about and managed to gain enough support to experiment. While the intention may be good, often those projects are not scoped to production, or they don't have the right people involved."

Deborah's advice to transform your organization is to tackle your most important challenges and/or opportunities. "Be bold, go big, just be careful to start small and use experimentation as part of an agile process." To illustrate this point further, she points to an IBM-American Airlines case study (3,4), a digital transformation project that would deliver dynamic rebooking and other self-service digital tools to the airline's customers using IBM's technology platforms. "It's an amazing case study because of the enormity of the undertaking. There are some who would hear about a big idea like this and feel that it is too big an undertaking: 'This will take us years to do.' But this was a strategic goal of the company, and at every turn, the team was focused on how to make it happen. Anyone who needed to be involved went into a war room to figure out what it was going to take. The project was done in just four and a half months from the start of development to deployment to customers. This is an amazing story because if people feel like these projects are too big, they tend to shy away from them and settle for another, less important experiment, which is a shame because it's the big, bold ideas that have the power to transform companies."

A company's ability to get to AI at scale often comes down to C-level involvement. "I want leaders to be able to recite their strategy. And it needs to fit on one page, which is enough to get the organization behind the goal. They all need to agree, top management and line management alike, exactly what things they want to accomplish and the outcomes they want to achieve in the next 12 months."

Results

"You know, one of the things that I'm most proud of at IBM are the AI executive education sessions we conducted for senior managers. We got the right senior executives together in a room and taught them the basics of AI. In the sessions I was told, 'This is just what has been missing.' We need to demystify AI for our business leaders so they can have a meaningful contribution as they're the ones who are close to the business problems.

"In one of the sessions we tackled 'redefining innovation.' Every company has problems they just live with. Maybe people tried to fix them before and found they were unfixable, so companies just accepted them. Now we are getting our leaders to bring those back to the table. We have so much data and so much computing power, thanks to the cloud, we actually may be able to fix the things that the executives don't even talk about anymore.

"Part of the AI-for-executives sessions was to *rethink* the way you solve problems. The message to them was, think of the outcome you are trying to solve. That's where the most successful transformation is coming from. Once you open the door and you exercise that muscle used to think *outcomes*, it is likely you're ready to make real progress!"

Advice to Digital Leaders: Assure Fairness and Explainability

"Machine learning algorithms are easily tainted. Sometimes, an unconscious bias of the engineer writing the model creeps in, and sometimes bias comes from the training data with an unintended result. Take the Apple Card, for example. It all started when a tech executive realized that he was granted 20 times the credit limit of his wife, despite having joint assets and filing a joint tax return. Goldman Sachs, the issuer of the Apple Card, confirmed that gender is not, in fact, a data field included in the algorithm. Nonetheless, the model observed a historical pattern indicating that men are entitled to higher credit limits than women and was able to infer the bias.

"This sparked an investigation by the State of New York's Department of Financial Services, which investigated Goldman Sachs for violating the law and discriminating against women—'whether intentional or not.' (5) It was really eye-opening for many people to realize that the model can be biased, even if you purposefully exclude problematic fields. The fact that the model learned to be biased from training data was very eye-opening for many."

In addition, it is important to focus on explainability. "As we consider putting the machines to work, we have a responsibility to ensure fairness and to make sure that we can explain how the model arrives at a recommendation. Fairness has to be a conscious endeavor, and it's the responsibility of the C-level and the board to know that these are the models that will drive the next generation of what we do. They have to ask themselves, have we taken the right and prudent steps to make sure that they are indeed fair?"

Chapter References

1. Building the AI-Powered Organization. https://hbr.org/2019/07/building-the-ai-powered-organization.
2. AI Demands a New Perspective. https://medium.com/@deborah.leff/ai-demands-a-new-perspective-10976a3db843.

3. IBM Cloud Flies with American Airlines. https://www.youtube.com/watch?v=t1PgNr8VMLc

4. American Airlines: The route to customer experience transformation is through the cloud. https://www.ibm.com/case-studies/american-airlines.

5. Apple Card Investigated After Gender Discrimination Complaints. https://www.nytimes.com/2019/11/10/business/Apple-credit-card-investigation.html

CHAPTER 5.9

Be Flexible, Learn to Lead from the Front–and from Behind–When Transforming Large, Established Organizations

KRISHNA CHERIATH

VP, Head of Digital, Data, and Analytics, Zoetis Inc., and Former CDO, Bristol Myers Squibb

Profile

Krishna Cheriath recently took on the leadership of Digital, Data, and Analytics at Zoetis, Inc., a global animal health company. Zoetis is the world's largest producer of medicine and vaccinations for pets and livestock. Krishna is responsible for driving an effective digital, data, and analytics strategy.

Prior to this, he was the Chief Data Officer for Bristol Myers Squibb (BMS), a global biopharmaceutical company. He served as BMS's data strategist for improving data quality and as BMS "evangelist" for data sharing and data-driven innovation. He was also considered a lead technologist for data and analytic automation.

Krishna has over 27 years of experience in digital, technology, data, and analytics. Before Bristol Myers Squibb, he was a management consultant working for such firms as TCS, PWC, IBM, and Accenture. Krishna received his undergraduate degree in electrical and electronics engineering from the University of Kerala and his MBA from the Stern School of Business at New York University.

Krishna is a thought leader. From his experiences as a management consultant advising fortune 500 companies on digital strategy and technology acceleration to being an adjunct professor at Carnegie Mellon University, Krishna is in a position to offer learned insights on the success of digital strategies—as well as their leaders.

Key Theme: Accordion Strategy Allows Us to Flex

The Job of a CDO Is Multidimensional

We first discussed Krishna's multiple roles as Chief Data Officer. "My role is multidimensional—it is part data strategist, part digital strategist, part data evangelist, part technologist, part data shrink, and part empathizer around the quality of the data. I help connect the company's digital strategy to its business imperatives to ensure that the data availability, velocity, quality, and control are paramount, so the organization is able to derive value from its data.

"I also describe my role as something similar to an accordion. When I meet CDOs from other industries, most of them say, 'You will have a tenure of two, maybe three years. There is a lot of turnover.' You will have a honeymoon period when you have political capital, but if you spend this period trying to do an 'empire-building' exercise or trying to launch into a big transformational program that doesn't yield value for multiple years, you will not succeed. You need to be able to come in, help craft a vision, then be able to do two things at the same time. One is making sure that you are adding value in the here and now and that you are solving real business questions. The second is to simultaneously invest in the infrastructure, i.e., the plumbing, to reimagine the digital and data ecosystem for the organization's long-term sustainability.

"If we talk about R&D and their very gifted computational researchers, they may not need help in algorithm development, nor do they need help in machine-learning modeling—but they may need help in making sure they have access to the right data. You need to have that dexterity to be able to effect the right digital outcome. I partner with business leaders, analytic leaders, and technology leaders to craft an enterprise vision around data. We have to make sure that the internally generated, as well as externally acquired data, is at the right level of availability, velocity, quality, and control. We also have to ensure that our stakeholders outside the company—our patients, physicians, and other healthcare providers—look at us as a trusted digital partner. They need to be sure that we acquire, collect, manage, and use data responsibly and ethically."

> "We will flex our enterprise services to match with the consumers' needs and abilities. This is especially important when you're implementing an enterprise-wide strategy—in some cases you lead from the front, and in some cases you lead from behind."

In Krishna's view, the accordion approach has helped build trust across the organization. "It takes a village—you need a good working coalition of business leaders, technology leaders, and analytic leaders to travel in the same boat as you. The accordion concept is a way for us to say, we will flex our enterprise services to match with the consumers' needs and abilities. This is especially important when you're implementing an enterprise-wide strategy—in some cases you lead from the front, and in some cases you lead from behind.

"If you focus on achieving the right *enterprise* outcome, and if you exhibit a certain level of flexibility in your approach, you will enlist the support of your stakeholder group. Your peers will perceive what you are trying to accomplish, and your ideas will gain support. That is the only way I know to build organizational influence and to sustain it."

Boundary Spanners Work Across Business Functions

We questioned Krishna on the particular challenges he faces transforming large, established enterprises. "Enterprises of significant size and scale have a very wide spectrum of digital savviness. In both healthcare and life sciences, when you look at the stakeholder groups you see advanced R&D teams with best-in-class scientific and computational researchers. And then you see other parts of the organization that may not have that same depth of digital talent. So how do you make sure you have the 'right size strategies' to enable and power those communities that are really deep into data science, and at the same time support those functions and communities that don't have the benefit of this knowledge?

"The second challenge is what I call the 'unicorn' challenge. You need people who understand the business strategy—the ones who are conversant in the business vocabulary and can talk business strategy with the business folks. They are able to reflect on how digital can be a key lever to advance the business strategy, then turn around and convert that knowledge into a coherent execution plan. These people need to be *boundary spanners,* who work seamlessly across business functions.

"If I imagined this as a physical office, the first is the business leader's office, the next is the analytic leader's office, the next is the IT leader's office. The boundary spanner I describe is able to go into each office and engage in a conversation around the leaders' various strategies and convert them into an integrated business, analytic, data, and technology plan. That's why I call it a unicorn combination. If you are going to be in the digital space, and you want to make traction in digital strategies, you need that kind of talent and leadership."

Digital Strategy and Company Vision

"Companies need to have strategic clarity. What is the core company strategy, and how can the digital strategy be a lever to accelerate and augment that strategy? This is a very important clarification because it is easy for digital to become a *distraction*. For life sciences companies that are focused on discovering transformational medicines for their patients, digital can help by accelerating the speed at which we bring them to our patients. With all the hype around digital and AI, and everybody trying to become the next Netflix or Airbnb, you can easily lose strategic focus and clarity. The company strategy drives digital strategy, which in turn drives data, analytics, technology, and talent strategies."

Ethics in AI—Particularly in the Field of Medicine

We also asked Krishna to elaborate on ethics in AI related to the healthcare industry. How can we tell the story where AI is augmenting human work, not replacing human work? In response, Krishna pointed us to an article he wrote for the pharmaceutical industry, "From Hype to Health: Delivering the Promise of AI in Biopharma," where he advocates for a multidimensional strategy and explains how you can have both responsible and innovative AI. (1) "In this case, it is not an either-or equation. You need to make sure that you define the right use cases where AI could augment their outcome. I'm a big believer in the augmentation strategy that requires us to invest in several layers. One is identifying the right opportunities where AI can be of value. Then how do you have the right experimentation mindset so it can prove its value? This can generate a lot of pilots and proofs of concept. As a friend of mine said, 'There are more AI pilots than there are pilots at United and Delta airlines combined.' I thought it was a good line because there's so much money going down this path. But there is not enough investment around *sustaining* AI and the lifecycle management of AI—in other words, the care and feeding of AI.

"The other aspect of this equation is talent development, which needs to be viewed from two different angles. One is the talent needed for scaling and adopting AI. It is a highly competitive segment, and the nascency of AI makes it tough to get the right talent. The other is enabling the knowledge workers—the larger employee population—to *coexist* in an AI-augmented enterprise. They should not see AI as a threat or competition but see it as a way to capitalize fully on a business opportunity."

Building Trust in AI

We asked Krishna how he builds trust in the organization when employees may be worried about AI and other machine-learning technologies.

"I think about trust and transparency very broadly. If you look at society at large, there are many examples of where there is a lack of faith in institutions. People have a hard time agreeing on facts. You can see many aspects of the fraying of digital trust in society. Companies need to think about trust and transparency, not just as a check-the-box exercise—you have to go beyond that. I think companies in the future will compete on trust and transparency, which is a different mindset. You're now saying that you want to be the most trusted company by your customers in this space; that means that you will have to do things differently.

> "The larger topic of trust and transparency requires you to examine how you are engaging with your stakeholder base and reflect on how you're collecting the data, how you're using it, if you have the right systems in place, how you train your people, and how you are equipping every employee to be a good *digital citizen.*"

"AI is one aspect of it, but the larger topic of trust and transparency requires you to examine how you are engaging with your stakeholder base and reflect on how you're collecting the data, how you're using it, if you have the right systems in place, how you train your people, and how you are equipping every employee to be a good *digital citizen.* There's a lot more that needs to be done in training and educating people so they can apply digital technologies to their jobs. That's why I think about trust and transparency in a broad way, to make sure the organization has an end to end view of it."

Results

Krishna has seen the results of digital initiatives. The core tenets of the digital strategy are materializing, and tangible results are being realized in companies he has worked with. Krishna emphasizes that "Companies that embrace an intentional digital strategy, including a cloud-first strategy, will achieve nimbleness and an agility of response when new business opportunities are identified."

Advice to Other Digital Leaders: Trust and Transparency Are Key in the Digital Age

"The first step is to have strategic clarity with the *why* around digital transformation. Following a me-too approach based on hype is never a recipe for success. It requires you to invest time and energy to have the right alliances to make sure that you craft a digital strategy that makes sense for your company and fully capitalizes on the company's strengths.

"Second, anybody who is signing up to be a digital leader is in effect signing on to be a *first among equals*. You have to have a lot of dexterity to navigate many different stakeholders. You need to be an *alliance builder*, you need to be a community builder, and you need to know when to lead from the front and when to lead from behind. Approaching it from this mindset allows you to bring together the right set of stakeholders. Digital strategies are not going to be successful on a standalone basis. They have to have strong connections to the business strategy. You need to have the CIO in your corner, you need to have a business leader in your corner, you have to have the CFO in your corner. And, of course, you need to have your CEO support it."

> "Digital strategies are not going to be successful on a standalone basis. They have to have strong connections to the business strategy. You need to have the CIO in your corner, you need to have a business leader in your corner, you have to have the CFO in your corner. And of course, you need to have the CEO supporting it."

Chapter References

1.   Cheriath, K. (2020) "From Hype to Health: Delivering on the Promise of AI in Bio-pharma." *Pharma Boardroom*, https://pharmaboardroom.com/articles/from-hype-to-health-delivering-on-the-promise-of-ai-in-biopharma.

CHAPTER 5.10

Listening Is the Key for Transformation

DOMINIK SCHLICHT
CEO, Talbot New Energy AG

Profile

Dominik Schlicht is the CEO of Talbot New Energy AG. The company was founded in 2011 by the Talbot family with the aim of developing technologies in the field of energy efficiency and to build and sell the resulting products. The Talbot STP—Steam to Power—is an innovative, production-ready energy efficiency system that uses industrial-process steam in the low-pressure range. Talbot New Energy offers the profitable generation of green electrical energy, improves the energy balance of customer companies, and sustainably relieves the burden on the environment.

Prior to his role as CEO, Dominik was the Head of Engineering, Product Optimization, and Special Affairs at ABB Power Grids (now Hitachi ABB Power Grids), driving the digital, lean transformation for High Voltage Services, restructuring and leading the technical department, and transforming technical know-how into strategic key success factors. Dominik was also the Head of Aftermarket Business at Voith Turbo, a specialized company for intelligent drive technology in Essen, Germany, where he was involved in the analysis, optimization, and digitalization of service processes.

Key Theme: Listening Is the Key for Transformation

"In the end, only the result counts. In order to achieve those results, we need to connect with strategic partners and build a trusting relationships between suppliers and customers. We never let our customers down. To do that, we have to connect customer needs to our employees' perceptions and build mutual trust. We have to listen carefully to employees, integrate their feedback, appreciate their work, and observe and manage their fears. This is the basis for the success of digital transformation."

Digital Strategy and Company Vision

Talbot's vision is to achieve maximum customer satisfaction by all possible means. Dominik firmly believes providing unique services will differentiate European companies like Talbot from international competition. "Investments in digitalization must always result in an increase in the performance of the service offering and improved delivery reliability. The use of data plays a decisive role in this respect."

Dominik points out that access to and evaluation of data across the entire value chain is an essential component in the digital transformation of manufacturers of capital goods. "On the supplier side, data is crucial for building a fully integrated supply chain and enabling forecasts of inventories and delivery times. On the customer side, data is becoming increasingly important to provide condition monitoring and to automatically incorporate information from the product lifecycle into product improvement."

Connect with Your Partners and Build Trusting Relationships

Dominik talks about the value of trusting relationship between suppliers and customers. "Trust has created a much greater willingness to provide relevant

146

data and information to each other. This enables a more holistic system perspective, which helps us build technical systems that are needed to increase customer service and product quality." This type of information-sharing can drive product optimizations with tangible customer benefit. "For example, from one partner we received all the data about their transmission, which was extremely valuable for us. And from an engine manufacturer, we received their complete engine data. We were then able to combine these data and optimize our clutch design."

Furthermore, Dominik talks about the importance of introducing key suppliers to the corporate strategy. "Our suppliers always get the relevant areas of our strategy explained to them. We consider our suppliers as strategic partners; we provide them information to reduce their uncertainty. We actually exchange forecasts with them."

Never Let the Customer Down

"Both ABB and Voith have great company values." A central mission statement that Dominik remembers, was the slogan, "We never let our customers down." He says, "If our customer has a problem, then it is our problem, too, and we have to solve it for them. It's good that we get money for it afterwards, but first, we must solve their problem. This basic perspective has helped us develop our 'service management circles,' i.e., regular meetings of our service leaders, where we discuss how we can deliver added value for the customer. What really hurts the customer right now? How else can we help them?

"In the end, only the results really count! The customer is not at all interested in the structure of your organization and internal processes. If we are organized differently—in an upside-down pyramid, for example—but the results are right, then the customer doesn't really care. What interests the customer is what they ordered, if the delivery arrives on time, and whether the product meets their expectations.

"It is also important to connect customer needs to your employee's perceptions in order to build mutual trust." In order to do that, Dominik uses extensive storytelling with a clear focus on the *why* and the *what*. "I try to draw the big picture for the people first. At Voith I told them, 'Imagine that you are a Russian rail customer who will soon invest $250 million in our company.' So people could empathize with the customer, we created a persona for them. Look, this is the client, this is how they look, this is how they act, and this is what they value. This helped our experts feel increasingly connected to the client and how we would market to them.

"In the second step, we created user stories. We tried to understand the user as well as possible, in an intuitive way. We tried to develop user stories by asking each other as many questions as possible about their business, its business context, possible problem situations, and likely technical solutions. We proceeded this way in a very close-meshed manner. As leaders, we coordinated this activity very closely, and when we had developed a mutual understanding of the user, we could see that mutual trust had developed at the same time. This made the whole topic of creating connections with the user much easier to handle."

Honesty and a Positive Attitude Are the Most Important Core Values

When asked which of the basic character values in the Trust Model is most important for a digital leader, Dominik answered *honesty*. "Especially in today's complex communication situations, honesty is the most important character. The more distributed we work and the more complex are the tools and techniques we use to perform our work, the more difficult it is to assess whether someone is honest with me or not—and the more cautious I am in my actions. If I am cautious in my actions because I am afraid the ice I'm standing on will crack, then I will usually lose a lot of speed. And if I lose speed, then I lose innovation and progress."

Furthermore, Dominik believes that for a leader, attitude is particularly important. "The most important thing you do is stand in front, and people notice: 'Hey, he is present, he is here, he has a positive attitude, and he wants to achieve something.'" Dominik sees a person's qualifications merely as an entry ticket into the professional world. "It is like reading a book and then getting a certificate for training—anyone can do that. Qualification and training can now be found on the Internet in any form you like. Qualification is like an entry ticket to the job and the business world. That's all it is."

Integrate Employee Feedback into Your Communication Strategy

Dominik quickly realized that two-hour presentations on strategic topics once a quarter was not the way to reach employees and communicate the strategic impetus of the digital transformation efforts. "We noticed that the uncertainty in our market was increasing, and our employees tended to be rather insecure after these meetings. The employees were unable to identify with many topics, the buy-in was very low, and the company itself was generally in a very un-certain situation. We realized that we simply had to communicate better with our people and address the fears of the employees. Instead, monthly meetings were introduced, presenting only one major topic at a time, allowing time for employees to ask questions and for us to listen to them closely. We made these questions and answers available to the employees via SharePoint so that even those who were not present could read through everything."

In addition to the established monthly meetings, Dominik introduced the Objective Key Results, or OKR, a methodology with daily, weekly, and monthly targets. At the monthly meetings, employees had to present the prog-ress of their OKR. "At these monthly meetings, we had our employees present their project for five to ten minutes. This meant I didn't have to go through all the reports and perhaps forget what was presented. The employees really appre-ciated giving their presentations—they could now express their ideas and status to 100 people at one time. This format also had unanticipated effects. When you see that your colleague has presented an interesting project, you tend to walk up to him and say, 'Listen, I have a useful idea regarding your project.'"

For Dominik, one of the main reasons that digital transformations fail is that employees—and their fears—are not taken seriously and are only super-ficially considered. "I have an employee who has been with the company for 40 years, who had to be out for medical reasons. When he came back to work, he saw that the seating arrangement was different. His landline phone was gone; he had to dial in differently than usual because a new process had been introduced that he was not familiar with. He is an exceptional expert in what he does. He had optimized his personal system, but we had, unintentionally, destroyed his coral reef and built a highway over it!

"We are a technology company that builds on the know-how of its employees. If you forget this and don't always take the employees with you, then you will lose. Even small changes that do not seem relevant at first can sometimes be a disruptive factor for employees."

Connect Digital Initiatives to the Core Processes of Your Company

Dominik has observed what it is like to be a CDO who has to manage an international company—and what can go wrong. Under the CDO, many projects were initiated, but from Dominik's perspective, "the position served more as a marketing metric rather than driving a profound change. The CDO came from Silicon Valley, of course. They had probably done a lot of great things, but only at very high, strategic levels. We had a period in which 180 digital solutions were launched with all kinds of facets. However, no matter who I talked to, no matter which division, no one could remember if any of the solutions were in use by a customer. In other words, it was all a great marketing joke, and board members were certainly talking a lot about how great it all was, yet nobody could show me a real project where it was implemented. The CDO was ultimately dismissed and responsibility was given to the individual divisions."

Dominik sums it up, "When you start these digital initiatives, you have to link them to the core know-how of your business processes, for example the production processes. If you can't do that, then you have a great marketing cannon, but it's like a hot-air balloon. At some point it bursts, and nothing is left except hot air."

Results

"Trust in the organization and in the cooperation with suppliers and customers is the basis for the success of a digital transformation." Dominik impressively describes how, in his role as a leader, he has built trust in cooperation with partners, in cross-functional teams and, how, in his opinion, this is a *prerequisite* for being able to work in a goal- and solution-oriented manner. While working at Voith, Dominik, together with supply-chain partners, managed to reduce order-processing time from six to eight weeks to 48 hours. Compared to other supply chains, this created a clear and unique selling point. "Achieving this objective was extremely important to building trust in our partner network. Satisfying the *ultimate customer need* is the most important thing for these organizations."

According to Dominik, honesty is the decisive characteristic of a leader. As a result, companies such as ABB and VOITH can realize very demanding digitization projects in order to ultimately achieve their vision of a 100% customer-oriented company and thus secure their internationally competitive positions.

Advice to the Leaders: Network Outside Your Interest Group

"I always suggest that companies operate in networks. To operate in networks means to be in regular exchange with different interest groups. Try to avoid your own interest groups if possible. It can be an exchange of information about the university via your alumni network, or participation with a Chamber of Industry and Commerce. It might be sponsoring a project on a research advisory board, or it might entail working in a professional association. It could even be participating in virtual groups on LinkedIn, where you can get new inspirations on a daily basis.

"I would really put 'extending my network' on your calendar. It is a great way to get new inspirations about digital transformation and to talk about your own vision and roadmap and to receive feedback and reflections on them."

CHAPTER 5.11

Digital Accelerates and Magnifies Traditional Metrics of Running a Business

CRAIG MELROSE

Executive Vice President, Digital Transformation Solutions, PTC

Profile

Craig Melrose is the Executive Vice President of Digital Transformation Solutions at PTC. He leads an organization that builds customer-facing solutions using industry-leading Computer Aided Design (CAD), Product Lifecycle Management (PLM), Internet of Things (IoT), and Augmented Reality (AR) technologies. Craig is a thought leader in operations excellence and digital transformation and an expert on large-scale Industry 4.0 programs. Prior to joining PTC, Craig led numerous operations and digital transformation initiatives during his 20-year career at McKinsey & Company. Throughout his career, he has helped companies enhance their factory automation strategies, including Toyota Motor Manufacturing, where for five years prior to joining McKinsey, he led the improvement of Toyota's lean production systems through the introduction of new products.

PTC (NASDAQ: PTC) is a global software and services company that empowers digital transformation for industrial, aerospace and defense, electronics, and other companies through industry-leading 3D (CAD), PLM, IoT, and AR solutions.

Key Theme: Digital Accelerates and Magnifies Traditional Metrics of Running a Business

"Companies have always had a P&L statement and a balance sheet. I think the financials of running a business is the native language of business—it's been around for a thousand years and it will be around for a thousand more. Companies have always driven improvements of how they operate and how they perform, whether it's cost reductions, revenue growth, or some combination of the two. Now you have digital technology available at your fingertips, which is enhancing, reinforcing, and accelerating all the traditional improvements of business.

"Digital adds speed to achieving those traditional financial metrics—it's got double-digit impact and is really transformational. When I think transformation, I define it as a caterpillar-to-butterfly' transition; once you're a butterfly, you can't go back to being a caterpillar. The performance is such that you would never want to *undo* this transformation, give it back, or lose it because it generates double-digit millions of profit and double-digit percentages of margins. It's that level of magnitude, it is that type of speed, in about a two- to three-year window."

Craig underscores that digital transformation has to be at the enterprise level to reach absolute double-digit impact. "When you're getting to that order of magnitude, it is truly business-changing, the benefits of which can be reinvested and utilized in a multitude of ways. It becomes a competitive advantage. Digital unleashes speed and scale to the traditional metrics of improvement."

Digital Strategy and Company Vision

"PTC is both a digital business and a technology company—they're intimately interwoven; it's impossible to disaggregate one from the other. Because we are on the tip of the spear in terms of providing solutions and products to our customers to drive their digital transformations, we have to be aware of our own digital transformation and digital strategy. If we're not living it, we can't help others to live it; this interwoven nature makes it impossible to separate. The strategy for our company, and for any company, is to continue to grow, to continue to be diversified, to continue to strengthen what already is internal to the company and even expand our offerings, either through organic growth or acquisition.

"So that's our strategy; digital is at the foundation of our digital strategy because it's a digital offering, it's a digital product. It's something that will help somebody else deliver a digital experience that allows them to achieve a higher level of performance. They're hand in hand."

PTC's Digital Transformation Journey

We asked Craig how far along PTC is in its own digital transformation journey.

"In my viewpoint, I can look back, and I can't see the starting line, and I can look forward, and I can't see the finish line. I think it's a bit like when you're learning to swim, and your parent or the instructor keeps backing up, and the finish line keeps getting farther and farther away. This helps you strengthen those muscles and learn. I don't know that our finish line is achievable. And if it is, it's time to find a new race, because otherwise we'll stagnate. I think PTC is far enough along the journey, but it's a path that may never end. That's okay. As long as we create more experiences, more successes, accomplish more

> "As long as we create more experiences, more successes, accomplish more goals, it's a positive journey down the path of digital—for both ourselves and our customers."

goals, it's a positive journey down the path of digital—for both ourselves and our customers."

Build Trusting Relationships

"My experiences are that people are created for relationships, but there's always a social, interactive dynamic to people—and to me. Trust is the currency of relationships that are powerful, successful, and symbiotic. If I can truly trust someone, I'm able to divulge more to them, and they're able to help me on a level

> "Trust truly unlocks those relationships, which then unlocks the power of the individual, the power of the team, the power of the company, and you are able to transcend what would be normally just a transactional relationship."

that I'm giving them access to. Trust truly unlocks those relationships, which then unlocks the power of the individual, the power of the team, the power of the company, and you are able to transcend what would be normally just a transactional relationship.

"It is an 'I know you, let's accomplish item A. Okay, we've accomplished item A, thank you' type of relationship versus 'How are you doing? How's the family, how's the dog, how can I be helpful to you, how can I help you navigate a personal or professional issue? This is my aspiration, what do you think? Give me feedback, give me advice.' When you're taking it to that level of relationship, you're able to transcend and achieve a different level of performance. Teams and individuals are willing to put themselves out there to be trustworthy and to earn trust. This two-way, omni-directional interaction around trust creates the ability to achieve a higher level of performance."

Trust Is Necessary in a Cross-Functional Team

Craig is a big proponent of building trusting relationships while role modeling and helping his team understand the value, meaning, and intent of such relationships to achieve more and to take performance to a higher level.

"My team is cutting across all product and functional teams to take bits and pieces from each and create a solution that didn't exist before. Due to the cross-functional nature of the digital transformation solutions team at PTC, we need to build trust, put our arm around someone, and bring them along.

"There are three ways to get something done. It can be *done to you, done for you,* or *done with you. To* and *for* are failure modes; *with* is the only right way to proceed. So how do we do this *with* that individual—what is their motivation? What is their need of accomplishment? How do we define their success and our success? We need to do it *with* them, not *to* them or *for* them. So due to the nature of the cross-functional influence required, trust and building relationships is a huge component of my team. By extension, we're role modeling that to the rest of the organization in those interactions."

Building Trust with Sphere of Influence

The digital solutions that Craig is spearheading at PTC are transformational and can have a profound impact on any company. We asked Craig, "How do you build client relationships so they can achieve the full benefit of PTC's solutions? "

"I describe these relationships as a sphere of influence. If it's with a general manager or a plant manager of a facility, their sphere of influence is their facility. That is a smaller sphere of influence within an enterprise, but it may be large enough to be the *toehold* of a digital transformation for that company. They may also work with a mid-level manager—they may be on the IT side or on the operations side. Eventually, both need to be involved. The IT team is going to integrate the solution with all the existing systems and understand its architecture, and the operations team needs to understand its advantages and benefits.

"A different sphere of influence would be with regional or country leaders, business unit leaders, or at the product leadership level of a company. That's a broader sphere of influence; they may have multiple facilities or multiple products in the field. They may focus on all of them, or they may focus on just one. They are at a slightly higher level of transformation within a company, but probably not a high enough sphere to drive an entire enterprise transformation. A senior leadership team certainly has the sphere of influence to drive it across the entire organization. Any starting point is fine. Irrespective of where you start, speed and scale come into play. If you start at a more senior level, it's easier to go faster and get to that higher level of scale. I think organizations are starting to see that more and more."

Craig reiterates that trust and trust-based relationships are important, regardless of the sphere of influence. You have to have a trust-based relationship to be truly transformative, because you're putting yourself out there—and you're expecting others to put themselves out there, because of what's at stake on a personal, professional, or corporate level."

Results

PTC has successfully transitioned to a highly recurring subscriptions revenue model that accounted for 94% of total software revenue in 2019. "PTC is growing year over year in the low double digits and has for a while; that is the proof of our success. Our solutions have helped customers achieve some impressive results as well. It's not always revenue and growth—they may be changing their cost structure or their operating model. They may be improving their service levels, decreasing lead times or inventory levels, or even reconfiguring a network.

"We've also seen various degrees of success; some are just beginning and may not be realizing the full potential. Some are well down the path and are realizing the full potential, while others are so far down the path that they're actually setting new records. It is all positive improvement because of leveraging digital."

The Trust Equation

Craig talks passionately about "the Trust Equation" that McKinsey has used and he has embraced. (1) The equation is straightforward:

Trustworthiness = Credibility + Reliability + Intimacy (or Integrity) / Self-Orientation

"*Credibility* speaks to what we say and our credentials. *Reliability* is how others perceive the consistency of our actions and how our actions connect with our words. *Intimacy* is how secure, or safe, others feel in sharing information with us. *Self-Orientation* refers to our personal or self-focus.

"So you apply numbers from one to ten to each dimension of the formula. There's no such thing as a zero; you have to have a positive score in each dimension. So the maximum trustworthiness score would be ten for each dimension in the numerator, for a total score of thirty. You then divide this by a perfect *Self-Orientation* score of one. This results in a maximum trustworthiness score of thirty."

"If your trust score is between twenty and thirty, I feel good about the situation, but I'm always aspiring to be a thirty out of thirty. To get a score in the low twenties, I need to be a seven or an eight on all three dimensions, which is approximately an 80% effort or better to be *credible*, an 80% effort to be seen as *reliable*, and an 80% effort to demonstrate more *intimacy* or *integrity*. I'm putting myself out there and being vulnerable. My team members also put themselves out there and are being vulnerable; this really helps build relationships with our customers. It's a high bar—we have to be proficient, across the board, in the numerator.

"Now for the denominator, if my *self-orientation* is not a one, there's no way I will achieve the trustworthiness I desire. Even a thirty divided by two is fifteen. If there's the least bit of 'it's about me,' or worse yet, our customers start saying 'I don't trust him—I think he's too concerned about how he's appearing, so he's not really paying attention,' trustworthiness is destroyed.

Advice to Other Digital Leaders: Never Be Self-Oriented; Release Your Agenda

"Never be self oriented; release your agenda. Truly put yourself in other people's shoes, truly understand their motivations. Understand how making them wildly successful will also make you wildly successful. Remember, it's all about them being wildly successful and you coming along for the ride. That was the biggest *Aha!* learning of my entire career, which is why I so embrace that formula and live by trust-based relationships. All this is foundational, which I think always needs to be in place. Now take that foundation and add digital to it.

"Digital, in some ways, is still being understood, still being embraced. Since it's not well defined, not universally understood, not embraced fully, it can be apprehensive, scary, or nerve-wracking. So there's an opportunity to become self-oriented, which prevents the trust from happening.

"There lies the biggest opportunity to build trust in digital. Satisfied clients say, 'You've helped me to be proficient at it; you've helped me to actually prove it to the rest of my company. Now I'm the towering leader inside my company, shining a light on the path to digital transformation. You've helped me to do that; you're my partner in this. You put me in the limelight and allowed me to do this; now I am going to embrace it.' When the clients say this, we've just built a wonderful, lifelong trust-based relationship on a topic that is still at various levels of understanding, proficiency, and confidence in individuals and organizations alike."

Chapter References

o   David H. Maister, Charles H. Green, and Robert M. Galford, *The Trusted Advisor*, New York: Free Press, 2000

CHAPTER 5.12

Make Ideas Happen–Openness, Honesty, and Innovation Drive Success

DAGMAR WIRTZ

CEO, 3WIN

Profile

Dagmar Wirtz is the CEO of 3WIN. Since founding the company in 1999, she has received multiple awards for her visionary management style. She is considered a model entrepreneur by Germany's Federal Ministry of Economics, Family, and Youth. Dagmar served as a founding member of a think tank that helped regional companies adopt digital skills. German magazine *Stern* awarded Dagmar for her socially oriented way of managing employees and the way in which she and 3WIN handled challenging economic times in the industry. *Stern* also gave Dagmar top marks for 3WIN's progress in digital transformation and changing the culture and working style.

3WIN is a family business located in Aachen, Germany. The company manufactures individual high-precision parts, small assemblies, and complete machines using latest technologies and AGILE production methods. 3WIN relies on its cooperation with highly ranked universities to keep up with the latest developments in the field of mechanical engineering. Customers consider 3WIN a creative partner in the production and optimization of individual parts, assemblies, and turnkey solutions.

Key Theme: Make Ideas Happen—Openness, Honesty, and Innovation

Dagmar promotes openness, honesty, and innovation throughout her company. She is willing to approve almost any project if employees had good reasons to start the initiative and can explain its potential value for 3WIN—even if she personally had not considered it before. She sees innovative experimentation as a "hobby" of her company, affectionately calling it her "craft room." This open-minded, innovation-based approach is a core part of Dagmar's strategy in leading her medium-size family business. She believes that innovation in mechanical engineering can only happen when a concept is implemented; turning an engineering model into reality is the only way to learn and innovate. Dagmar's reliance on this ambidextrous approach and her collaborative projects with industry, universities, and institutes have positioned 3WIN as a prime example of trustworthy mechanical engineering that is "Made in Germany."

Digital Strategy and Company Vision

3WIN focuses on open manufacturing methods by engaging in research projects with universities and institutes as well as sophisticated mechanical engineering products with industrial customers. Both fields involve complex tasks and generate a high proportion of 3WIN's innovations. Targeted communication and the concept of "co-development" are fundamental features of 3WIN's strategy. The company claims to be a *Machbarmacher*, which is best translated into English as "making the possible happen." This requires a highly flexible organization that utilizes cross-sectional personnel to create a high degree of in-house vertical integration and value for their clients. Digitalization processes, for example digital workflows, are used to accelerate innovation, increase transparency, and simplify communication.

Connect Your Employees to Digital

Dagmar points out that "making the possible happen requires specific structural and cultural elements." She observes that digital transformation, which seems to be an omnipresent topic, is less developed in small and medium-sized companies. To overcome this problem, Dagmar founded a think tank and a digital media workshop four years ago to show how the work of the future will look different. "The central focus of the paradigm shift in digital transformation—especially in manufacturing companies—is the person who, as a company employee, is constantly confronting new challenges arising from new technologies. Demonstrating these new technologies in our digital media workshops is especially important to me in order to keep connected, not only to my employees but also to have a meaningful level of conversation with our partners. The topics surrounding digitization are incredibly exciting and will move us forward. But we have to break it down so that the worker on the shop floor understands what's happening and sees it as an opportunity to keep his or her job."

For Dagmar, a crucial element of leadership is the attitude and mindset of a leader towards their tasks and responsibilities. "It's not their knowledge or qualifications. If you don't want to drive a task forward, if you don't want to make a company successful, then it's irrelevant whether or not you have the qualifications or the knowledge. There are people who are incredibly qualified. I know such people; they have a wealth of knowledge, and I enjoy being around them, but if there is no one who drives the engine over and over again—and motivates everyone—then none of this is of any use." She fondly remembers, "My father used to say, if you have to carry a hunting dog to hunt, then it's not a hunting dog."

Demand Openness, Transparency, and Honesty

Dagmar values openess and transparency, and they are evident in her leadership at 3WIN. "What my employees do in their free time is their business, but as far as the company is concerned, I expect absolute transparency. We have created work tools and an information plattform accessible to all employees, where all project and engineering information is stored transparently. It is very important to me that everyone is not cooking their own little soup. In addition, I encourage employees to speak their minds and share their opinions. I say what what fits and what doesn't fit. And I expect the same from my employees. I don't like apple-shiners who don't say what they think—it doesn't help us."

This kind of openness and transparency allows 3WIN to stay on top of their game. "Usually there are two to three idea providers who always come up with something new. Then I provide a certain time frame, a certain budget, and then I

> "If you don't respond to new ideas, then you don't motivate anyone to bring in any ideas."

say, 'Do it!' That's how we get along at the end of the day. If you don't respond to new ideas, then you don't motivate anyone to bring in any ideas.

"However, at a certain point, a clear line must be drawn. We have to accept that trying things in such a pragmatic way might end up in a failure as well. Accepting errors is part of the innovation culture at 3WIN, and making the right conclusions is part of the business culture. If an idea comes along and I think it could be appropriate to create a new business—and if it fits our core capabilities in any way—then we will do it."

During the COVID-19 crisis, for example, 3WIN developed a contactless disinfectant dispenser, then took this product and modified it for a totally different purpose. "Many people fear they may get infected when attending church, so we developed a *contactless holy water* dispenser that will be installed in churches and, of course, we designed it in a very noble style. Now we have to see if it will be a success for us."

Building Trust in Crisis: Be Agile and Take Advantage of Your (Sometimes Hidden) Competencies

As an entrepreneur, Dagmar has faced several crises, but she was able to navigate them successfully, thanks to her agility. "I just become totally agile—now I can think of hundred thousand things that might sustain us." In 2006, when the economic crisis hit her company, 3WIN was in the midst of expanding its locations. Due to the crisis, customer orders stopped from one day to the next. But Dagmar managed to keep all employees on board. "We rounded up the team and talked openly. We have this new property, and there is actually a lot to do. Now, who can do what to complete the building?" 3WIN actually rebuilt the entire facility with most of the same people from the company. Some of them could lay tiles, others could do drywall construction and roofing. "There were also electricians in our company, so our employees could do all kinds of things beyond their previous field of work. Of course, this also creates trust when you manage something like this together. These are things you don't forget." This effort was recognized outside of the company as well. German magazine *Stern* awarded Dagmar and 3WIN for the ways they handled challenging economic times in the industry.

Results

For a small German enterprise in the manufacturing industry, staying successful over the years is a great challenge. Even for special-purpose machinery, price pressure from international competition is high, and the only strategy that counts is the strategy of *differentiation through unique offerings.* "Digitalization is only one means to stay competitive. The real challenge of machine manufacturers is to truly become unique in offerings which requires a combination of unique attitudes towards customers, unique behaviors of making things happen, and unique skills."

3WIN is successful in shaping this specific culture, which is a result of Dagmar's leadership principles and style. Her leadership efforts are concentrated on making 3WIN an admired workplace. As a result, the company has been repeatedly awarded for its achievements. 3WIN topped the list of "most innovative medium-sized companies in Germany" by *Stern* magazine. It was also recognized as one of the "top ten manufacturing companies in the Aachen region" and received the "Made in Aachen" Seal of Approval. Dagmar herself has been recognized as a model entrepreneur by the "WOMEN enterprise" initiative.

Advice to the Leaders: Be Genuinely Interested in Your Employees

Be genuinely interested in your employees, that's my advice. "My father said to me, 'You really are concerned for your people. I never behaved that way to the extent you do.' Dad, nobody used to do this in the past. But today you have to be interested in the people and their needs. I want to have a good atmosphere and a good *tone* in my company. I am here 100 hours a week, from morning to evening. I don't want to have any disgruntled people around me. I want to have people who feel good, who like to do their job, and who are willing to stand up for something. This is very important to me, and a good understanding of our relationships is a major part of it."

CHAPTER 5.13

AI-Powered Enterprises Require Visual Analytics

RAHUL C. BASOLE

Managing Director and Global Lead for
Visual Data Science, Accenture AI

Profile

Dr. Rahul Basole is Managing Director and Global Lead for Visual Data Science at Accenture, focusing on developing and delivering new competencies at the intersection of visual analytics, data science, AI, and strategy. His expertise includes advancing and applying novel, interactive, and human-centered visual analytic approaches to understanding, designing, and managing complex processes, enterprises, and ecosystems and bringing effective data-driven visual solutions to the C-suite.

Rahul is a globally recognized thought leader in visual data science and strategy and his award-winning research has been published in leading management science, computing, and engineering journals and conferences. In his prior roles, he has served as Professor in the School of Interactive Computing at Georgia Tech, Director of the Institute for People & Technology, and a Visiting Scholar at Stanford University. Rahul earned a BS in industrial and systems engineering from Virginia Tech, an MS in industrial and operations management from the University of Michigan, and a PhD in industrial and systems engineering from the Georgia Institute of Technology.

Accenture is a global professional services company with leading capabilities in digital, cloud, and security. Accenture AI takes human ingenuity and artificial intelligence and applies them at the core of business to help clients become intelligent enterprises and solve their most complex business problems.

Key Theme: AI-Powered Enterprises Require Visual Analytics

Rahul says, "Data is ubiquitous—we use it, we create it, and we act on it in all of our daily economic and social activities. Yet, at the complexity, speed, and scale at which data grows and is consumed, it is increasingly difficult to make sense of it. This is particularly the case as we are rapidly entering the age of analytics, automation, and AI.

"Visual analytics—the fusion of analytical models with interactive visualization—promises to be a crucial link in connecting humans with complex data, analytics, and AI, and people are starting to really embrace it. That's why I was brought to Accenture. There is a growing understanding that organizations are stitching data together, building models, and running analytics on top of it, but at some point, someone may have to consume it. The human-data interface is increasingly important, and that's become a real big push for many firms across industries.

The field of visual analytics is still fairly new. We're not talking about static BI reports created with off-the-shelf data-visualization systems. We're truly thinking about how to integrate AI and visual analytics to understand and explore complex organizational contexts and support operational and strategic processes through intelligent multimodal interfaces. In order to transform the potential into reality and accelerate from proof-of-concept to proof-of-value and scaled production, we need new skills, we need new capabilities, and we need digital assets and accelerators. This requires the building of trust within and outside of the organization. And that's a massive challenge."

A Novel Application of the Digital Twin Concept—Business Management

We asked Rahul if he would please give us an example of how visual analytics might be applied at a leading customer's company.

"There are lots of interesting applications of visual analytics across many industries and enterprise functions. One fascinating example is that of the digital twin idea *applied to business management.* As you know, the core concept of a digital twin is that it is a digital replica of some physical object, system, organization, people, or process that exists (or could potentially exist). Now let's apply the digital twin concept to business management. Many organizations use a massive set of visual dashboards to capture and communicate the state of specific parts of their enterprise (think supply chain or marketing or HR or finance). More Digitally Mature organizations may integrate these across functions, but that's not always the case. Unquestionably, such application areas fill an important role and act like 'control towers,' but they often fail to realize the true potential of interactive visual analytics in business, for various reasons, as these dashboards serve relatively simplistic, sometimes static, and often merely one-way communication or reporting purposes.

"With the transition to a truly integrated digital backbone, near or even real-time data, and massive cloud computing power, organizations are starting to embrace more advanced analytic capabilities that enable them not to observe current or historic aspects of their enterprise, but also to proactively manage the future state of their enterprise more dynamically. To do that, organizations need to think of digital twins more as 'flight simulators'—a platform that creates a representation of both the structure of the system, its many interconnected entities, including people, processes, and organizations, and the overall ecosystem it is embedded in, as well as its complex behaviors and strategies. With this more advanced version, organizations can now play out, experiment with, and visually observe strategies, futures, and scenarios. The power of this approach is that alternate business models can be evaluated, evolutionary paths can be played forward and backward, and costly and perhaps even highly risky decisions can be evaluated before actually implementing them. By putting the human decision-maker at the center of this highly interactive visual analytics experience, significant new business value can be created.

"This example highlights the tremendous potential of visual analytics, in particular where humans need to interface with highly varied data and complex contexts for decision-making. This is particularly important right now when times are so uncertain for many organizations. 'Where are the greatest risks for my supply chain? Who are my customers, and what are they demanding? What are my competitors doing? How should we position ourselves in our ecosystem?' Organizations need to be able to truly immerse themselves in these issues, drive their decision-making with data, and *see* them come to life. We believe that these *interactive* visual analytic systems personalized to the stakeholder can offer tremendous communication, sensemaking, discovery, and decision-making capabilities."

For a better understanding of the state of visual analytics and its likely impact on business, please see the article Rahul published titled "Visualization 4.0: The Renewed Relevance of Visualization for Business." (1)

How to Scale Visual Analytics

"While data visualization applications are plentiful in organizations today, visual analytics as an enterprise capability is still in its infancy. As with most emerging, unproven, or unknown digital solutions, you often have to start small to show what they can really do and enable organizations to achieve, perhaps do things differently than they have ever been done before. Unless you want to get stuck in the proof-of-concept valley, that's the challenge. Customers ultimately want scalable solutions, and we need solutions that can be rapidly deployed and proven.

"We—and by that, I mean both the practitioner and academic community broadly—are very good at developing visual analytics proofs of concept. In academia, that was the primary focus—focus on small data sets, i.e., small, well-defined problems, etc. But if I have to create a digital cockpit for an enterprise, I have to shift from small proof of concept to complex proof of value at scale very rapidly. I need to consider large sets of data, many stakeholders, and established work processes. Moreover, I have to gain organizational trust by convincing decision makers that I'm going to give them something that traditional reporting and dashboard tools could not do before.

"It helps when you have a forward-looking or visionary client. If you have a client who says, 'I understand data and analytics, and AI is important to unlock value for me,' it really helps! Most of the CXO's I talk to definitely think that way and are fully aware of the potential. But are they willing to embrace something relatively new and perhaps not scalable across their entire enterprise? Not everyone will. Some organizations are starting to lead and separate from the pack by embracing such digital innovations. Others are lagging. The key is to show them quick wins that have value. Doing that, you will gain their trust. If you can then scale the solution and accelerate towards industrialization, the buy-in really blossoms."

Dealing with Complexity

"Every organization has a fairly diverse portfolio of systems and solutions that it leverages to create and provide value for its internal and external stakeholders. Digitally immature organizations may still operate legacy infrastructures and siloed applications. For many modernized organizations, this portfolio can include things like cloud infrastructures, Internet of Things, blockchain, cybersecurity, platforms, a sophisticated activation layer, and so on.

"Ultimately, it's a set of digital technologies that you have to assemble and orchestrate, sometimes managed internally, sometimes by external partners and vendors. Put together and managed correctly, they can create tremendous value. I thus believe that digital transformation is ultimately *complexity management*. Organizations must manage highly intertwined infrastructures, orchestrate a multiplicity of vendors, and consider many solutions. There is no silver bullet digital strategy; one-size does not work, but there are some proven approaches. You simply have to embrace and manage against this complexity.

"Why talk about complexity when considering visual analytics in an AI-powered enterprise? Because the majority thinks of visualization and visual analytics as merely consumption endpoints. Is

> "There is no silver bullet strategy; one-size does not work, but there are some proven approaches. You simply have to embrace and manage this complexity."

this the data dead end? Far from it. I posit that highly successful visual analytic solutions are fully embedded in business processes and workflows, ingesting data from various systems, enabling insight generation that in turn activates decisions and tasks. Visual analytics should thus be viewed as an important element of the overall digital portfolio. Moreover, while we hope that a single off-the-shelf solution may power our visual analytics, today it is more likely composed of a set of tools and libraries. Understanding this inherent complexity is thus crucial."

Understand the Client's Business Needs

"Transitioning to the cloud and creating the right digital backbone can almost be considered table stakes now. What can start differentiating organizations from one another is how data, analytics, and AI are applied to solve their problems, serve their customers, and create value. To do that, we have to think deeply about what a CXO may need in terms of solutions. We have to think about their domains, the context that they are in, the industry they are in, and the forces that are shaping their industry. It's the combination of understanding the problem-data-user-context that leads to the most successful solutions. This is no different for visual analytics.

"One of the strengths of Accenture is that we have deep experience across many industries along with a massive bench with tremendous skillsets that can mobilize quickly. We can leverage what we've learned in one industry and transfer it to others. Being able to translate that knowledge across domains is very important. Our clients trust us with our ability to think deeply about their problem-data-user-context, devise feasible near- and long-term strategies, design solutions, and ultimately help them execute and implement them. Having that end-to-end capability is extremely powerful."

Understanding the Value at Stake

"Embracing change is critical. You have to make people uncomfortable, even if they think they are currently in a leading market position. I often lead off conversations by making organizations aware of the diverse external forces at play—that can fundamentally disrupt what they are doing—no matter how comfortable they might feel. That's what I call the *value at stake.* This is a really important concept in the digital transformation space. If you don't understand the value that's at stake, to continuously think about the future and transform, you will likely be unable to actually capture that value and adopt new solutions. Your organization may be on a great trajectory, but transformative forces can appear at any time and totally disrupt your business model. By building trust through demonstrating value, you can help them navigate the changing business environment. You cannot run without sometimes crawling and walking first. So starting small, producing quick wins, and then showing these clients the value of their decisions is my formula for success."

> "Your organization may be on a great trajectory, but transformative forces can appear at any time and totally disrupt your business model. By building trust through demonstrating value, you can help them navigate the changing business environment."

Accenture as a Trusted Partner Is a Big Help

"It helps to have a brand that is trusted by many companies. Accenture is trusted by clients and partners across the globe and ecosystem. We partner with major cloud providers, we work with major data and analytics providers, we find and nurture new startups. We believe that to truly bring value to our customers, we need to understand what and who is best-in-breed in the world. The shiniest object is not going to solve their problem—it's the right solution that they need built for them. Our ecosystem of partners is tremendously important to achieving this, in particular in a new field like visual analytics."

Results

"Visual analytics is still a novel concept to many organizations, even within my organization, and it will take time to be broadly adopted. We are starting to build momentum by forming cross-disciplinary design, development, and execution teams, creating accelerators, identifying partners, and planning for centers of excellence. Right now, cloud, analytics, automation, and AI is on everyone's mind. Addressing the human-data interface through visual analytics is a natural next step.

"Visual analytics is going to shift from nice to have to absolutely essential for the AI-powered enterprise. But to get them to the point where it becomes almost second nature will, I think, still going to take some time. We need to move away from the notion that visualization is a one-way medium of reporting and communication. Visual analytics is more than that; it's a capability that allows you to explore and discover contexts, make the complex transparent and explicable, challenge assumptions, and understand, compare, and evaluate interventions and strategy alternatives that you haven't thought about.

"With the growth of the algorithmic enterprise, there is this notion that humans will be superfluous. I believe the contrary. While algorithms may replace certain processes and tasks, humans are still quite critical, and the ability to understand how algorithms and AI are designed and implemented, identify and mitigate biases, and augment decision-making can only be amplified through visual analytics."

Advice to Digital Leaders: You Need to Be a Systems Thinker and Storyteller

"We are in an exciting time for digital leaders. Challenges and opportunities are massive. To lead successfully, I think Covey's core values—integrity, intent, capabilities, and results—provide an important framework. To me, the four values are highly interdependent. You can't really have one without the others. If you're good at one, that's great, but you need to be decent at all four."

"The most successful leaders that I've seen have an innate ability to connect the dots—to be systems thinkers—and can communicate extremely effectively to their team and to their clients.

"I spoke about complexity and value at stake earlier. Unless you understand how things fit together and why, both on the problem and solution side, you will not be able to truly embrace complexity. That's why being a systems thinker, one who can see the parts and the overall system, is critical.

"At the same time, it's not just about understanding the problem and perhaps the solution but also about communicating both effectively. A digital leader thus also needs to be a good storyteller. Visual analytics can in fact be part of your storytelling toolkit.

"You also have to have a vision moving forward. If you can balance the near term with the long term, I think that makes for a really successful digital leader. Some of the folks who have taken me under their tutelage have really done this for me. I'm learning from them on a daily basis. Hopefully, I'm giving them something with visual analytics for AI-powered enterprises that they haven't thought about before."

Chapter References

1.   Basole, R.C. (November-December 2019) "Visualization 4.0: The Renewed Relevance of Visualization for Business." *IEEE Computer Graphics and Applications*: 39(6), 8-16.

CHAPTER 6

THE WINNING FORMULA

"There is no alternative to digital transformation. Visionary companies will carve out new strategic options for themselves—those that don't adapt, will fail."

– Jeff Bezos

Here is where it all comes together. Let's take a hard look at what these digital leaders have told us. What key themes have they employed? What are the key actions they have taken to build trust in their organization? Lastly, what advice do they give you to build confidence in your own organization?

Before we get into these actions, let's step back and see what general observations we learned from them about the field of digital leadership.

General Observations

- Digital strategy and business strategy are inseparable. They are intertwined

- Many positions can provide digital leadership

- It takes a village

- Trust-enabling actions reduce cost and accelerate speed of transformation

- Digital is a force multiplier; it amplifies traditional metrics

- Digital transformation is complexity management

- Digital leaders are boundary spanners

- 'Cloud first' is the predominant digital strategy

- Despite all the ballyhoo, AI still is in its infant development stage

- The turnover of CDOs is very high

- People are the real key to digital transformation

Figure 31. *General Observations*

Digital Strategy and the Business Strategy Are Inseparable; They Are Intertwined

Seth Kaufman (Ch 5.7) said, "It's not about having a social media strategy or digital strategy; it's about having a business and people strategy, of which digital is a huge piece. Any successful organization with a sustainable business strategy should have digital embedded throughout."

Craig Melrose (Ch 5.11) added, "PTC is both a digital business and a technology company—they're intimately interwoven; it's impossible to just disaggregate one from the other. Because we are on the tip of the spear in terms of providing solutions and products to our customers to drive their digital transformations, we have to be aware of our own digital transformation and digital strategy. If we're not living it, we can't help others to live it—this interwoven nature makes it impossible to separate."

And Krishna Cheriath (Ch 5.9) summed up his organization's digital strategy when he said, "It is actually seen as business critical. Companies need to have strategic clarity. What is the core company strategy, and how can digital strategy be a lever to accelerate and augment that strategy? This is a very important clarification, because it is easy for digital to become a distraction. Digital strategies are not going to be successful on a standalone basis. They have to have strong connections to the business strategy."

Many Positions Can Provide Digital Leadership

At the beginning of the book, we asserted that digital leadership is not defined by a person's rank or position but rather their ability to earn trust (see Introduction). Evidence of this is the list of proven digital leaders we interviewed—seven CEOs, two business unit heads developing digital products, two functional heads leading corporate functions, and two CDOs. All of them have contributed to the successful digital transformation of their organizations by addressing their very beliefs—and *will*—to change.

Marc Schlichtner (Ch 5.6) takes a very pragmatic view. "Everybody involved in such a transformation needs to have the same mental model and mindset." Marc realizes digital transformation is hard work, and to change the beliefs of the organization will require effort, focus, and determination. Chuck Sykes (Ch 5.1) picks up on this idea of changing beliefs: "Over the years, one of the things that I've always shared first and foremost with people is yes, I am the CEO, and that conveys a certain amount of responsibility that I have, but I am not the only leader in the company. I am a leader among leaders, and I always tell my leaders, don't ever lose that perspective."

Mindsets and mental models guide your perception and behavior. They are the "thinking tools" that you use to understand life, make decisions, and solve problems. Learning a new mental model gives you a new way to see the world.

It Takes a Village

None of our digital leaders work alone—or even suggested that they do. They all depend on partners to implement digital transformation. This partnership can range from external business partners to other leaders in the company. The phrase "it takes a village" is only part of the African proverb, "It takes a village *to raise a child.*" This means that an entire community of people must interact with the children for them to experience and grow in a safe and healthy environment. What does this proverb have to do with digital transformation?

In a 2019 *Harvard Business Review (HBR)* article titled "Digital Transformation Is Not About Technology," the authors urged organizations to "leverage insiders"—in other words, people from the same village. The article added, "Organizations that seek transformations (digital and otherwise) frequently bring in an army of outside consultants who tend to apply one-size-fits-all solutions in the name of 'best practices.'" They recommend the best way to transform your organizations is to "rely instead on insiders—staff who have intimate knowledge about what works and what doesn't in their daily operations." (1) Krishna Cheriath (Ch 5.9) states that to build trust in the organization, "It takes a village—you need a good working coalition of business leaders, technology leaders, and analytic leaders to travel in the same boat as you."

Trust-Enabling Actions Reduce Cost and Accelerate Speed of Transformation

All digital leaders claimed that they witnessed dramatic improvement in the performance of their organizations, sometimes doubling it. They all related this to the actions they had taken to enable and build trust within the organization (See "The Winning Formula: Fifteen Key Actions" below). This is a powerful endorsement of Covey's Speed of Trust thesis and our hypotheses and the essence of our book. We will go into these actions and examples in more depth later.

For now, accept that we are definitely seeing the benefit of these actions. A world-class culture enables your people to perform at their highest levels, increasing productivity and efficiency. Because personnel costs are typically fixed costs (you're paying someone the same wage whether they're working at 60% of their own capacity or 100%), any amount of increased productivity goes right to your bottom line.

Trust Is a Force Multiplier

Craig Melrose (Ch 5.11) achieves this synergy across his organization through the relationships he creates that are built on trust. "Trust is the currency of relationships that are powerful, successful, and symbiotic. If I can truly trust someone, I'm able to divulge more to them, and they're able to help me on a level that I'm giving them access to. Trust truly unlocks those relationships, which then unlocks the power of the individual, the power of the team, the power of the company, and you truly are able to transcend what would be normally just a transactional relationship."

Digital Transformation Is Complexity Management

Rahul Basole (Ch 5.13) and other digital leaders described the job of a digital transformation as complexity management. "Organizations must manage highly intertwined infrastructures, orchestrate a multiplicity of vendors, and consider many solutions. There is no silver-bullet digital strategy; one-size does not work, but there are some proven approaches. You simply have to embrace and manage against this complexity."

McKinsey and Company, in their article, "How do I manage the complexity in my organization?" point out that "carrying on as normal won't make complexity go away." They go on to say, "Organizations that report high levels of complexity are poor at creating value. This is not an issue that top management can afford to ignore or delegate to others. Leaving employees to struggle with complexity in the hope that they will eventually learn how to manage it wastes energy and resources. When people find it hard to get things done or decisions made, their morale suffers—and frustration sets in. To prevent this downward spiral taking hold, the top team needs to address the issues head on." (2)

If complexity is widespread, that might mean clarifying accountabilities and processes throughout the whole organization. However, when done right, digital transformation allows you to simplify complex processes and focus on what's necessary for your business.

CDOs Are Boundary Spanners

Closely related to the observation that digital transformation is complexity management is the observation that CDOs must be boundary spanners. In addition to the significant technical competencies they possess, CDOs need the ability—and patience—to work across business functions to resolve these complexities.

Krishna Cheriath (Ch 5.9) describes the people he needs: "You need people who understand the business strategy—the ones who are conversant in the business vocabulary and are able to talk business strategy with the business folks. They are able to reflect on how digital can be a key lever for advancing the business strategy, then turn around and convert that knowledge into a coherent execution plan. These people need to be boundary spanners, who work seamlessly across business functions."

"Cloud First" Is the Predominant Digital Strategy

There is no single digital technology—IoT, Digital Twins, or AI for example—that deliver both speed and innovation. The best combination of technology and tools for a given organization will vary from one vision to another. Yet every respondent claimed the *first* digital technology they utilized was the cloud. In fact, it is considered the "ante" to get in the digital transformation game.

The reasons for this are obvious. Even if your digital strategy is based on AI, you still need a powerful, on-demand computing capability that can be quickly configured for your applications. You don't want to get caught up—certainly at this point in your transformation—in a lengthy reconfiguration or purchase cycle. If you want to scale fast, you have to use the cloud. Knowing about cloud and cloud services is the prerequisite for how a business model can scale. It allows you to design data-based business ecosystems. It's foundational for everything else.

Krishna Cheriath (Ch 5.9) sums it up, "Companies that embrace an intentional digital strategy, including a cloud-first strategy, will achieve nimbleness and an agility of response when new business opportunities are identified."

Despite All the Ballyhoo, AI Is Still in Its Infant Development Stage

Discussing her company's digital transformation roll-out, Deborah Leff (Ch 5.8) observed a recurring theme: "Most executives thought they would be further along with their AI initiatives than they are by now. It seems that an inordinate number of projects have not been delivering the value they had expected—they either get stuck in experimentation or take significantly longer to put into production than they had anticipated."

In their article "Predictions 2020: AI Aspirations Will Both Simmer and Sizzle," Forrester says, "As with any megatrend, AI has captured the imagination and envy of many corporate leaders. Our data shows that many groups across the enterprise (software developers, B2C marketers, data and analytics decision makers, mobility decision makers, etc.) have already *tiptoed* into some form of AI." They predict, however, that 2020 will be different "when companies become laser-focused on AI value, leap out of experimentation mode, and ground themselves in reality to accelerate adoption." They further predict that CDOs and CIOs "who are serious about AI will come to the rescue, with a top-down mandate to get around the data access problem. *Firms with chief data officers (CDOs) are already about 1.5 times more likely to use AI, ML, and/ or deep learning for their insight initiatives than those without CDOs. Leadership matters.*" (3)

IBM offers another solution to stop "tiptoeing" around AI. In their article "A Radical Solution to Scale AI Technology," they conclude, "Most C-suite executives know they need to integrate AI capabilities to stay competitive, but too many of them fail to move beyond the proof of concept stage. They get stuck focusing on the wrong details or building a model to prove a point rather than solve a problem. That's concerning because, according to this research, three out of four executives believe that if they don't scale AI in the next five years, they risk going out of business entirely. To fix this, we offer a radical solution: *Kill the proof of concept. Go right to scale.*" IBM came to this solution by surveying 1500 C-suite executives. "While 84% know they need to scale AI across their businesses to achieve their strategic growth objectives, only 16% of them have actually moved beyond experimenting with AI. The companies in our research that were successfully implementing full-scale AI had all done one thing: they had abandoned proofs of concept. The companies that did this attempted scaling twice as often, succeeding at their scaling initiatives twice as often, and—because they were structured correctly and could incorporate what they learned along the way—ended up not only completing scaling projects more quickly but spending less money on pilots and fully scaled deployments.

The result? They achieved nearly three times the return on their AI investments when compared to their lower-performing counterparts. When you consider that *the average company in our study spent $215 million on AI in the past three years, the 54% difference represents a $115-million gap in missed returns from AI. And it's not just about the money. Successful scalers report significant benefits in customer service and satisfaction to workforce productivity to how efficiently the companies utilize their assets.*" (4)

Deborah Leff (Ch 5.8) and other digital leaders could not agree more. So, what's holding companies back? When we asked Deborah how companies should break out of experimenting with AI and go into full scale production, she promptly replied, "These companies must: *1) focus on outcomes, 2) involve the right people, and 3) align AI initiatives to strategic objectives.*"

The Turnover of CDOs Is Very High

The turnover of CDOs is particularly concerning to us. Our research has told us that the tenure of CDOs in organizations is approximately two years. This typically staff position is meant to provide deep technology know-how in data and analytics and guidance to top management teams.

We see their jobs much like the Chief Electricity Officer of yore when companies were switching from steam-engine power and connecting their plants to the new source of energy. Electricity Officers were to explain this newfangled invention and help top management understand how to leverage it in their operations. But like Electricity Officers, could CDOs' days be numbered?

Back in 2012, Gartner was predicting that 25% of businesses would have a CDO in their top team by 2015. Fast forward a year or so, and Forrester was questioning whether the CDO was a fad or the future, and earlier this year, they reported that just 1% of firms plan to hire a CDO in the next twelve months.

While we don't think CDOs are quite ready to be relegated to a fad status, there is certainly a lot of pressure on them, and their turnover is considerable. Some may be in over their heads, but we wonder, could the top management's expectations of them be too high?

People Are the Real Key to Digital Transformation

All of our digital leaders agree with this observation. Dominik Schlicht (Ch 5.10) says, "One of the main reasons why digital transformations fail is that employees—and their fears—are not taken seriously and are only superficially considered. We are a technology company that builds on the know-how of its employees. If you forget this and don't always take the employees with you, then you will lose."

Andera Gadeib (Ch 5.2) says, "It's all about choosing the right people for the company. It is even more important to bring in new experience and knowledge from outside. It's not just finding the best technically skilled people. New employees must certainly have a high affinity with what the company does, and in particular with the topics of digitization and its implementation. You should take the opportunity to find employees who are better than you are at what they do. This is the only way for a company to actually grow beyond itself."

Marc Schlichtner (Ch 5.6) adds, "Basically I'm trying to motivate people who are doing the right things by helping them translate their messages to management. At Healthineers, the transformation process is both a 'bottom up' and 'top down' process, and the best way is when they both take place at the same time. Grassroots revolution—this is the T-Club—and a top-down commitment; it's not just lip service. It is really intentional. Then you have both synergies coming together."

While the digital leaders endorse this observation, it comes with an ominous warning. A study conducted by the IBM Institute for Business Value shows that executives surveyed "plan to prioritize internal and operational capabilities such as workforce skills and flexibility—critical areas in order to jumpstart progress. Participating businesses are seeing more clearly the critical role people play in driving their ongoing transformation. Leaders surveyed called out organizational complexity, inadequate skills, and employee burnout as the biggest hurdles to overcome, both today and in the next two years."

Yet the study finds a *"significant disconnect"* in how effective leaders and employees believe companies have been in addressing these gaps. *"74% of executives surveyed believe they have been helping their employees learn the skills needed to work in a new way; just 38% of employees surveyed agree. 80% of executives surveyed say that they are supporting the physical and emotional health of their workforce, while just 46% of employees surveyed feel that support."* (5)

HBR reported, "Directors, CEOs, and senior executives found digital transformation risk to be their #1 concern in 2019. Yet 70% of all digital transformation initiatives do not reach their goals. Of the $1.3 trillion that was spent on digital transformation last year, it was estimated that $900 billion went to waste." (1)

Why do some digital transformation efforts succeed and others fail? It's because most digital technologies provide *possibilities* for efficiency gains and customer intimacy. But if people lack the right mindset—*and actions*—to change, and the current organizational practices are flawed, digital transformation will simply magnify those flaws.

The Winning Formula: Fifteen Key Actions

Now that you see the challenging conditions digital leaders face, here are the actions our successful leaders recommend to you. Hopefully, they will help you rethink your, and your organization's, approach to leading a digital transformation.

The Winning Formula: Fifteen Key Actions

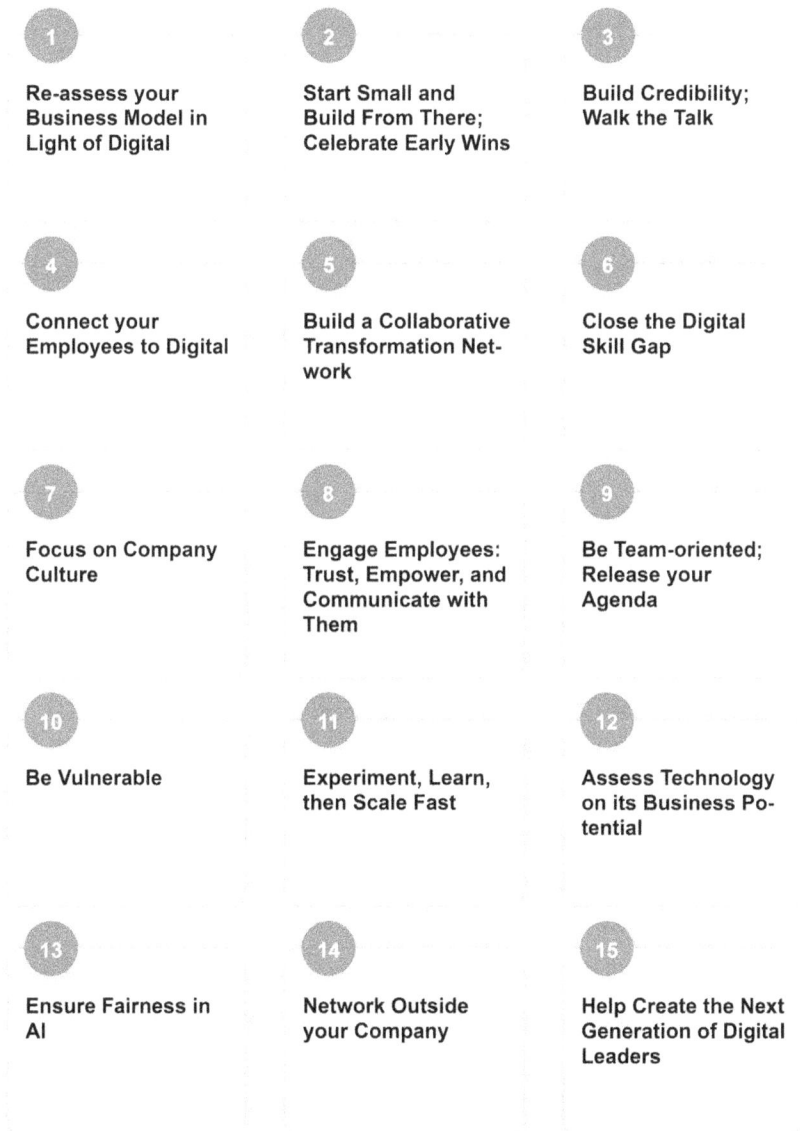

1 Re-assess your Business Model in Light of Digital

2 Start Small and Build From There; Celebrate Early Wins

3 Build Credibility; Walk the Talk

4 Connect your Employees to Digital

5 Build a Collaborative Transformation Network

6 Close the Digital Skill Gap

7 Focus on Company Culture

8 Engage Employees: Trust, Empower, and Communicate with Them

9 Be Team-oriented; Release your Agenda

10 Be Vulnerable

11 Experiment, Learn, then Scale Fast

12 Assess Technology on its Business Potential

13 Ensure Fairness in AI

14 Network Outside your Company

15 Help Create the Next Generation of Digital Leaders

Figure 32. *Winning Formula: Fifteen Key Actions*

1. Re-Assess Your Business Model in Light of Digital

Digital transformation, at its core, is basically thinking about how your business can use today's digital technologies to transform itself from being a physical asset-heavy company to an information-rich company. First start on the outside—check and scan the marketplace, then take an end-to-end view of your business and its operating models and assess whether the business model needs to change. If the answer is yes, then invest your time and resources to reinvent the organization's business model.

Please refer to Chuck Sykes (Ch 5.1), Larry Blue (Ch 5.3), Craig Melrose (Ch 5.11).

2. Start Small and Build from There; Celebrate Early Wins

Understand your goal for digital transformation and then focus on getting there. Starting small, especially during early stages of digital transformation, allows you to move fast. When you are a small team that is structured to work like a startup, you are more focused and can make decisions quickly. You don't have to deal with the broader issues around legacy processes. Nurture and protect projects during early stages with a visible and resolute commitment to their success. Celebrate early wins—people will respond to this; they want to be part of a successful team.

Please refer to Larry Blue (Ch 5.3), Rahul Basole (Ch 5.13).

3. Build Credibility; Walk the Talk

Building credibility is at the core of effective digital leadership. Credible leaders walk the talk, and we know that people learn by observing their leaders. If you do not have the trust of your people, and they take you for one who does not fulfill your commitments, you're going nowhere. You must be authentic, i.e., trustworthy, in everything you do—and by doing this, you will build trust-based relationships across the organization. These will serve you well.

Please refer to Larry Blue (Ch 5.3), Brandon Batten (Ch 5.5).

4. Connect Your Employees to Digital

You won't make it far without your people on board for this journey. So put your employees at the center of your digital strategy, align them with business goals, and clearly communicate your plan and vision for the company to them. The topics surrounding digital are incredibly exciting, but you have to break it down so that all employees understand what's happening, the reason for the change, and see it as an opportunity for their own personal success.

Please refer to Dagmar Wirtz (Ch 5.12), Dominik Schlicht (Ch 5.10).

5. Build a Collaborative Transformation Network

Don't go it alone. Develop digital acumen across the company by building collaborative networks inside the company, as well as with channel partners, customers, and suppliers. Evaluate existing cross-departmental relationships to see where innovations are occurring, where sufficient support is being provided, and where additional investments should be made. A senior leadership team certainly has the power and influence to drive transformation across the entire organization. However, it is important to cultivate other informal networks across the company, across all levels, to positively influence digital transformation initiatives. Furthermore, a good supportive network may just help you survive the early stages—often the weakest points—of the transformation journey.

Please refer to Marc Schlichtner (Ch 5.6), Craig Melrose (Ch 5.11).

6. **Close the Digital Skill Gap**

Get the right people on board. If you hire leaders who share your core values, then you are, in essence, establishing a leadership extension of yourself that will continue to perpetuate itself. Look for people who have real work experience with digital capabilities and methods, from the most basic to the most advanced skills. They must certainly have a high affinity with what the company does and understand the changes the digital strategy will have on the company's operations.

While it is likely necessary to bring in new experience and knowledge from outside, it is important to reskill and upskill your current employees to close the digital skill gap. Invest in training them on the skills needed for implementing digital operating models. Reallocate digital talent among business units and functions based on your strategy.

Please refer to Chuck Sykes (Ch 5.1), Andera Gadeib (Ch 5.2), Robert Kallenberg (Ch 5.4).

7. **Focus on Company Culture**

People are the real key to digital transformation. Company culture has to be transparent and inclusive. It should also promote an entrepreneurial mindset that is agile and risk-tolerant. You have to paint a picture of an exciting vision that people want to be part of, and then you have to provide them a path to get there. The technology is great, but as a leader, you need to educate and empower employees along the way. Seriously promote *continuous* innovation and the value of data. Apply data in your business processes and make the results visible to everyone to build a more "data-informed" organization.

Please refer to Chuck Sykes (Ch 5.1), Robert Kallenberg (Ch 5.4), Marc Schlichtner (Ch 5.6), Seth Kaufman (Ch 5.7).

8. **Engage Employees: Trust, Empower, and Communicate with Them**

Put yourself at the center of the transformation and deal with all facets of digitization. Make a constant effort to engage—listen to your employees, understand what they're thinking, and integrate employee feedback into your communications and operations strategy. Be sure to share success stories and the results of ongoing digital initiatives. When you respect the individual and nurture data-driven exploration, you will be on your way to building an innovative culture.

Please refer to Chuck Sykes (Ch 5.1), Andera Gadeib (5.2), Larry Blue (Ch 5.3), Dominik Schlicht (Ch 5.10), Seth Kaufman (5.7), Craig Melrose (Ch 5.11).

9. **Be Team-Oriented; Release Your Agenda**

Self-orientation—whether it is having a personal agenda, claiming credit, flattery, or embellishment—destroys trust. Never be self-oriented; release your personal agenda. Put yourself in your people's shoes and try to understand their motivations. Remember, making them wildly successful will also make you wildly successful. Again, in the end, it's all about *them* being wildly successful—and you coming along for the ride.

Please refer to Andera Gadeib (5.2), Craig Melrose (Ch 5.11).

10. **Be Vulnerable**

Being vulnerable means being relatable, truly caring, and genuinely interested in your employees. It also means being authentic and truthful, admitting mistakes, taking criticism —and doing it all in stride. When employees are willing to be vulnerable, they are no longer preoccupied with self-protection and defensive behaviors. This often helps build stronger relationships within and across your teams, which increases the flow of ideas and overall better results. Being vulnerable demonstrates the strength of character and emotional intelligence of the leader. Inarguably, digital transformation is one of the most technically demanding areas of business leadership, yet softer skills are proving to be the difference-maker.

Please refer to Craig Melrose (Ch 5.11), Rahul Basole (Ch 5.13).

11. Experiment, Learn, then Scale Fast

Experimentation is fine; it allows us to learn. However, it is important for leaders to have the skill to reflect and critique their decisions and then quickly adapt as necessary. Often, companies get stuck in experimentation. Without the interest and support to move it all the way though production, the experiment remains just an experiment. You must focus on outcomes, involve the right people, align digital initiatives to strategic objectives, and have the agility to scale fast.

Please refer to Brandon Batten (Ch 5.5), Deborah Leff (Ch 5.8), Krishna Cheriath (Ch 5.9), Rahul Basole (Ch 5.13).

12. Assess Technology on Its Business Potential

Evaluate technology—be it augmented reality, blockchain, internet of things, AI, ML, or whatever it might be—for its true business potential. A poor process automated is still a poor process—it is just automated, and no technology can help this. Focus on technology from a value and business perspective. Exercise caution: chasing the latest technology may take valuable time away from new processes or products with immense potential. Recognize if other companies are applying similar technologies in a much smarter way, which may disrupt your company.

Please refer to Brandon Batten (Ch 5.5), Dominik Schlicht (Ch 5.10), Craig Melrose (Ch 5.11).

13. Ensure Fairness in AI

Establishing trust, transparency, and ethics in AI are everything for its future. AI has many potential benefits for business, but these will only be realized if people trust these tools to produce unbiased results. Machine learning algorithms are easily tainted. Sometimes, the unconscious bias of the engineer writing the model creeps in; sometimes the bias comes from the training data—both cases, however, can create unintended results. Leaders have the responsibility to ensure fairness by explaining how the model arrives at a particular recommendation. Fairness has to be a conscious endeavor; have you taken prudent steps to make sure these algorithms are indeed fair?

Please refer to Deborah Leff (Ch 5.8), Krishna Cheriath (Ch 5.9), Rahul Basole (Ch 5.13).

14. Network Outside Your Company

The digital landscape is fast changing. As digital leaders, we have to understand the latest trends, take inspiration from what we read, see, and hear, and then think about ways to use it in the context of our companies. Make a personal effort to read; go to seminars on digital transformation, and lead by example—get the background and knowledge you need to make decisions that drive strategic investments. Network outside your company—Involve yourself in interest groups, attend university-sponsored programs, participate on advisory boards, or other avenues where you can get new inspirations for yourself and insights about digital transformation. Talk about your own digital vision and receive feedback and reflections on it.

Please refer to Brandon Batten (Ch 5.5), Dominik Schlicht (Ch 15.10).

15. Help Create the Next Generation of Digital Leaders

Several of the digital leaders we interviewed have been shaping digitization proactively by creating long-lasting impact through mentorship, community involvement, and helping to educate the new generation. They are giving back—whether it is championing a Girls Who Solve high-school program aimed at developing the next generation of women digital leaders, educating young people to find novel uses of digital technology in agriculture to make sure that the world will have the most sustainable and affordable food supply, helping shape the computer science curriculum in school classes, or introducing children to the development of software in a playful way by helping them build their first computer and creating their own first code. All these actions can go a long way in developing future employees for digital companies. Future digital leaders need a combination of technology, people, and entrepreneurial skills. And you can do your part to expose them to it.

Please refer to Andera Gadeib (Ch 5.2), Brandon Batten (Ch 5.5), Deborah Leff (Ch 5.8).

Call to Action

We explained the environment digital leaders work in. While we are sure it is very gratifying when they see the organization producing new results and significantly outperforming its former self, it is still full of challenges. Management's expectations may be set too high; not all employees may be able to be "reskilled" and fear the consequence of being laid off; and even new analytical technologies, like AI, are still full of hype and are being accused of generating biased results. And of course, there is competition. Every company, in every industry, is realizing the power and promise of transforming their organization and is feverishly trying to do so. Digital leadership is definitely not an easy row to hoe.

To help guide leaders on this journey, we captured the fifteen key actions the interviewees shared with us. We call this our "formula" for successful leaders to follow. Note that the formula contains a significant list of actions. (We never said this would be easy.) Also note that we call them actions and not beliefs or traits. These activities will require overt *action* on the part of the digital leader—there is no time to just "grow into them." (Please refer to figure 32, Winning Formula: 15 Key Actions.)

What Should You Do with This Winning Formula?

Our recommendation is to test yourself. There are two ways you can do this. You can simply rate yourself from 1 to 5 on each item, or you can have others in your organization *test you*. While the latter approach is definitely riskier, it does exhibit your willingness to reveal vulnerability. But, as we have repeatedly emphasized in this book, this says to your team, "I trust you—and I want you to trust me." This simple act is the first step towards making you a successful digital leader.

We would love to hear your story, your experiences in this transformation journey. Send us feedback by visiting our website, PatternsofDigitization.com. Who knows? You and your organization may be in the sequel to this book!

Chapter References

1. Tabrizi, B., Lam, E., Girard, K., & Irvin, V. (2019). "Digital Transformation Is Not About Technology." *Change Management, HBR.*
2. Heywood, S., Hillar, R., & Turnbull, D. (2010). "How do I manage the complexity in my organization?" *Insights into organization, McKinsey and Company.*
3. Sridharan, S. (2019). "Predictions 2020: AI Aspirations Will Both Simmer and Sizzle." *Forrester.*
4. Kanioura, A. & Lucini, F. (2020). "A Radical Solution to Scale AI Technology." *Product Development, HBR.*
5. IBM Newsroom (2020). IBM Study: Majority of Global C-Suite Executives are Rapidly Accelerating Digital Transformation due to COVID-19 Pandemic, but People and Talent are Key to Future Progress.

APPENDIX

INTERVIEW GUIDE

CHARACTER

INTEGRITY (Honesty, Humility, Courage, and Congruence)

1. (Honesty) How often do you publicly address employees' concerns with the strategy? Do you insist that employees be transparent and "talk straight" as well?
2. (Humility) How open are you to new ideas? What do you do when your employees come to you with ideas that may at first seem to be crazy, off-the-wall notions?
3. (Courage) Do you question the development paths you've set and take into account external, perhaps even controversial, perspectives?
4. (Courage) Can you give us an example of when you were willing to change your plans when these ideas proved beneficial? Do you give teams the time and resources to experiment with these new ideas?
5. (Congruence) Do you make decisions on intuition/experience or by utilizing facts and data analytics? Do you trust the data you are getting? What are the challenges?

INTENT (Motive, Behavior, and Agenda)

6. (Agenda) What is your digital strategy? What is your vision for the company? How does the digital strategy fit into the company vision and overall business strategy?
Note: Digital strategy may be embedded in the organization's overall strategy and focus on current operations (efficiency gains, automation) and/or new opportunities to capture value outside the current business (e.g., through new services or offerings).
7. (Behavior) How do you communicate your digital strategy to employees, suppliers, and customers? Did you do this personally or delegate it to your staff or the marketing department? Do you insist that your staff "communicate, communicate, and communicate?"
8. (Behavior) How did you (re)allocate resources to achieve the digital strategy? What specific investments did you authorize? Have you hired a Chief Digital Officer, formed partnerships with external organizations to help deliver the new strategy, launched a major acquisition and/or training program for data analysts, etc.? How much of your personal calendar is devoted to managing the transformation?
9. (Motive) Based on your judgement, do employees find the strategy exciting and believe it represents a real—and necessary—change to the current business? Can employees see themselves operating in this new model? How do you encourage their buy-in?

COMPETENCY

CAPABILITIES (Talents, Skills, Attitudes, Knowledge, and Style)

10. (Talents) What digital talents do you look for in your leadership team? Not all leaders are knowledgeable in digital transformation, much less

the information technologies that enable it. What is your recommendation on how to bridge this gap?

11. (Skills) What is your approach to hiring digital rock stars? "Digital natives" in specific departments may be necessary. Please describe how your organization recruits, hires, and retains people having these still rare skills.

12. (Knowledge) List the major technologies that underpin your digital strategy, e.g., Artificial Intelligence (AI), IoT, business process automation, blockchain, etc. Are you fully committed to the cloud? Have you gotten push-back from the IT organization? How did you handle this?

13. (Knowledge) Has your company institutionalized AGILE software development methods? To what extent are you leveraging AGILE, DevOps, and the cloud in your organization? What are the critical success factors on the way from traditional to AGILE project management?

14. (Attitudes) How easy is it to share the lessons learned across the organization? How do you enable and promote knowledge sharing? How keen is your company to learn from others? Does your company have "safety nets" to foster a culture of experimentation that embraces failure as an opportunity for learning?

15. (Style) Do you roll up your sleeves and work with the teams? How do you celebrate small successes and motivate employees?

RESULTS (Past, Present, Future)

16. (Past) What accomplishments do you believe led you to obtain your current position?

17. (Present) Thanks in large part to your leadership, we view your organization as a Digitally Mature organization. What recommendations would you give leaders of (less mature) Digitally Developing companies?

18. (Present) Digital transformation is hard because it is difficult to align the organization behind it. Moreover, as we have discussed, often the culture is not ready to embrace it. What advice would you give to someone in the midst of a digital transformation to address the people's fear of change?

19. (Future) What do you see as the next phase of digital transformation? How are you preparing your organization for what must seem to be almost continuous change?

FINAL QUESTION

20. In your mind, do the four principles of *integrity, intent, capability*, and *results* capture the essential elements of digital leadership, or is there a fifth principle that we have overlooked?

"In God we trust." Yes, and when we are at our best, we also trust in each other. Trust is fundamental, reciprocal and, ideally, pervasive. If it is present, anything is possible. If it is absent, nothing is possible. The best leaders trust their followers with the truth, and you know what happens as a result? Their followers trust them back. With that bond, they can do big, hard things together, changing the world for the better."

– George P. Shultz, former U.S. secretary of labor, treasury and state, and director of the Office of Management and Budget. Mr. Shultz is a distinguished fellow at Stanford University's Hoover Institution.

Source: Washington Post (2020). The 10 most important things I've learned about trust over my 100 years.

Notes

Notes

www.ingramcontent.com/pod-product-compliance
Lightning Source LLC
Chambersburg PA
CBHW072306210326
41519CB00057B/2812